**PERGAMON SERIES ON
ENVIRONMENTAL SCIENCE**
Series Editors: O. HUTZINGER & S. SAFE

VOLUME 2

SENSORY ASSESSMENT OF WATER QUALITY

Other Related Pergamon Titles of Interest

Books

* ALBAIGES: Analytical Techniques in Environmental Chemistry (Proceedings of the International Congress, Barcelona)

* FERGUSSON: Inorganic Chemistry: The Environment and Society

* HUTZINGER, VAN LELYVELD & ZOETEMAN: Aquatic Pollutants - Transformation and Biological Effects (Proceedings of the 2nd International Symposium on Aquatic Pollutants, Amsterdam)

   MOO-YOUNG & FARQUHAR: Waste Treatment and Utilization: Theory and Practice of Waste Management (Proceedings of the International Symposium held at the University of Waterloo)

   OUAND, LOTHANI & THANH: Water Pollution Control in Developing Countries (Proceedings of the International Conference held at Bangkok)

   SHAINBERG & OSTER: Quality of Irrigation Water

* VOWLES & CONNELL: Experiments in Environmental Chemistry

*In the Pergamon Series on Environmental Science

Journals †

PROGRESS IN WATER TECHNOLOGY (A Journal of the International Association on Water Pollution Research for the Rapid Publication of Conference Proceedings that Record Important New Advances and their Applications in all Fields of Water Pollution Control)

WATER RESEARCH (The Journal of the International Association on Water Pollution Research)

WATER SUPPLY AND MANAGEMENT

†Free specimen copy of any Pergamon journal available on request from your nearest Pergamon office.

SENSORY ASSESSMENT OF WATER QUALITY

By

B. C. J. ZOETEMAN

Head of the Chemical Biological Division,
Rijksinstituut voor drinkwatervoorziening
(National Institute for Water Supply)
The Netherlands

PERGAMON PRESS

OXFORD · NEW YORK · TORONTO · SYDNEY · PARIS · FRANKFURT

U.K.	Pergamon Press Ltd., Headington Hill Hall, Oxford OX3 0BW, England
U.S.A.	Pergamon Press Inc., Maxwell House, Fairview Park, Elmsford, New York 10523, U.S.A.
CANADA	Pergamon of Canada, Suite 104, 150 Consumers Road, Willowdale, Ontario M2J 1P9, Canada
AUSTRALIA	Pergamon Press (Aust.) Pty. Ltd., P.O. Box 544, Potts Point, N.S.W. 2011, Australia
FRANCE	Pergamon Press SARL, 24 rue des Ecoles, 75240 Paris, Cedex 05, France
FEDERAL REPUBLIC OF GERMANY	Pergamon Press GmbH, 6242 Kronberg-Taunus, Pferdstrasse 1, Federal Republic of Germany

Copyright © 1980 B. C. J. Zoeteman

All Rights Reserved. No part of this publication may be reproduced, stored in a retrieval system or transmitted in any form or by any means: electronic, electrostatic, magnetic tape, mechanical, photocopying, recording or otherwise, without permission in writing from the publishers.

First edition 1980

British Library Cataloguing in Publication Data
Zoeteman, B. C. J.
Sensory assessment of water quality. - (Pergamon series on environmental science; vol.2).
1. Water quality - Sensory evaluation
I. Title
628.1'61   TD380   79-41528
ISBN 0-08-023848-3

Printed in Great Britain by A. Wheaton & Co. Ltd., Exeter

To Adriana, Marianne and Margot

"Health is a state of complete physical, mental and social well-being and not merely the absence of disease and infirmity"

WHO, 1972, Basic Documents, 23 ed., 1

Contents

	Preface	xi
1	**Introduction**	1
1.1	General aspects	1
1.2	The nature of sensory assessment of water quality	2
	1.2.1 Types of senses for water quality assessment	2
	1.2.2 Human olfaction	4
	1.2.3 Human gustation	9
	1.2.4 Some effects related to chemoreception	11
1.3	Perceptible substances in drinking water	11
	1.3.1 Historical developments in The Netherlands	11
	1.3.2 Substances causing turbidity of water	13
	1.3.3 Substances causing colour of water	13
	1.3.4 Compounds causing odour of water	14
	1.3.5 Mineral substances causing taste of water	15
1.4	Outline of the work	16
2	**Sensory assessment of drinking water quality by the population of The Netherlands**	19
2.1	Introduction	19
2.2	The inquiry	20
	2.2.1 The questionnaire	20
	2.2.2 Preparation and realization	20
	2.2.3 Characteristics of the sample of the population	21
2.3	Sensory assessment of tapwater quality and possible causes of quality impairment	23
	2.3.1 Visual aspects	23
	2.3.2 Temperature	25
	2.3.3 Odour and taste	26
2.4	Aspects possibly affecting sensory water quality assessment	29
	2.4.1 Knowledge of raw water source and opinion on water taste elsewhere	29
	2.4.2 Odour of the air at the residence and at the place of work	30
2.5	Perceptible water quality aspects and general water quality assessment	30
	2.5.1 Acceptability rating of water quality	30
	2.5.2 Safety rating of water quality	32
2.6	Conclusions	32
3.	**Sensory assessment of 20 types of drinking water by a selected panel**	35
3.1	Introduction	35
3.2	The panel	35
	3.2.1 Panel selection	35
	3.2.2 Factors affecting odour sensitivity for o-dichlorobenzene and iso-borneol	38

3.3	Stimulus presentation, test conditions and psychophysical scaling	42
3.4	Selection and sampling of 20 types of drinking water	44
3.5	Taste and odour assessment of 20 types of drinking water	46
	3.5.1 Relation between taste and odour rating of drinking water	46
	3.5.2 Results of the taste rating	46
	3.5.3 Results of the taste quality assessment	48
	3.5.4 Factors affecting the taste rating by the panel members	49
3.6	Conclusions	51

4 Drinking water taste and inorganic constituents — 55

4.1	Introduction	55
4.2	Taste assessment of individual salt in water solutions by a selected panel	55
	4.2.1 Stimulus presentation and test conditions	55
	4.2.2 Results	56
	4.2.3 Discussion	56
4.3	Inorganic constituents in 20 types of drinking water and taste assessment	62
	4.3.1 Levels of inorganic constituents in drinking water	62
	4.3.2 Discussion	63
4.4	Conclusions	64

5 Drinking water taste and organic constituents — 67

5.1	Introduction	67
5.2	Analytical methods	67
	5.2.1 General remarks	67
	5.2.2 Organochlorine pesticides, cholinesterase inhibitors, polynuclear aromatic hydrocarbons and halogenated volatile compounds	67
	5.2.3 Organic compounds concentrated by means of closed-loop gas stripping and by means of amberlite-XAD adsorption	68
5.3	Detected types and levels of organic constituents in 20 types of drinking water	71
	5.3.1 General survey	71
	5.3.2 Compounds present in drinking water derived from ground water	73
	5.3.3 Compounds present in drinking water derived from surface water	74
	5.3.4 Comparison between tapwaters derived from surface water and those derived from ground water	76
5.4	Organic substances and drinking water taste	76
	5.4.1 Screening procedures to select taste impairing compounds	76
	5.4.2 Compounds with C/OTC ratios of 0.01 or more	78
	5.4.3 Some compounds lacking OTC references	79
5.5	Conclusions	80

6	**Sensory assessment of drinking water quality and health protection aspects**	**83**
6.1	Introduction	83
6.2	Sensory water quality assessment and water consumption	84
6.3	Sensory detectability and chronic toxic effects of compounds	86
6.4	Sensory detectability and acute toxic effects of compounds	86
	6.4.1 General aspects	86
	6.4.2 Natural and industrial compounds	87
6.5	Carcinogens and accompanying odorous compounds	89
6.6	Conclusions	90
7	**Further study and application of water quality assessment**	**93**
7.1	Introduction	93
7.2	Value of sensory water quality assessment	93
7.3	Application of sensory water quality assessment	94
	7.3.1 Relation between type of water and method of taste or odour intensity determination	94
	7.3.2 Determination of odour intensity of contaminated waters	94
	7.3.3 Determination of taste of drinking water	99
7.4	Measures to reduce perceptible contamination of water	104
	7.4.1 Causes of impaired taste and smell of water and fresh water organisms	104
	7.4.2 Prevention and reduction of taste problems within water supply systems	107
	7.4.3 Abatement of the presence of taste and odour impairing substances in surface waters	109
7.5	Final considerations	110

Summary	113
List of symbols, abbreviations and synonyms	117
Subject index	119
References	123
Appendices	131
2.1 Survey of z-values for associations between ordinal variables of the national inquiry	131
2.2 Survey of H-values for associations between ordinal and nominal variables of the national inquiry	131
2.3 Some results of the national inquiry, arranged according to 50 communities in combination with some water quality data	132

3.1 Survey of the characteristic data of the group used for selecting the panel members and their odour sensitivity for aqueous solutions of isoborneol and o-dichlorobenzene 134
3.2 Correction procedure for changes in the use of a taste rating scale by a panel 138
5.1 List of organic compounds identified in 20 tapwaters in The Netherlands in 1976 141
6.1 OTC/LD_{50} ratios for natural organic compounds in water 146
6.2 OTC/LD_{50} ratios for industrial organic compounds in water 147

Preface

This work is partially based on my thesis entitled "Sensory assessment and chemical composition of drinking water", which I defended on 28 April 1978 at the University of Utrecht, The Netherlands. This book, however, is given a broader scope as indicated by the title. Particularly in the last part of the work, additional information is incorporated giving a critical review of existing methods for sensory water quality assessment. Also methods are recommended with which the best results were obtained in our laboratory in measuring odour of contaminated water and taste of drinking water. Although many aspects of the problem of impaired taste and odour of water have been explored, new questions still remained unanswered. It is hoped that other scientists will find inspiration in this book to continue and to extend these water quality investigations and that as a result of these studies taste and odour assessment finds a similar broad application as was the case in the past centuries.

The studies presented in this work are of a multi-disciplinary nature. Their realization would therefore not have been possible without the active assistance of many others, some of whom I would like to mention: firstly Prof. Dr. E. P. Köster of the University of Utrecht, Prof. Dr. O. Hutzinger and Prof. Dr. R. L. Zielhuis of the University of Amsterdam, as well as Mr. J. G. de Graan and Mr. G. J. Piet of the National Institute for Water Supply, Voorburg, Mr. J. J. C. Karres of the Central Bureau of Statistics, Voorburg, and Dr. J. Kroeze and Dr. P. H. Punter of the University of Utrecht for their generous help and valuable advice. Also many other collaborators from the Psychological Laboratory of the University of Utrecht and the Chemical Biological Division of the National Institute for Water Supply have contributed significantly to this work by means of the organization of panel sessions, carrying out of chemical analyses as well as data handling. I am particularly grateful for the typing, preparation of drawings and correction of the text to Mrs. M.A.E. Wamsteker-van Houdt, Mr. D. Mos, Mr. H. van Lelyveld and my wife. The research discussed in this work would not have been possible without the active and stimulating support of Mr. P. Santema, Director of the National Institute for Water Supply.

I hope that this book, which deals with a still neglected but essential aspect of water quality, will find its way to all those active in the field of water supply and water quality control.

Maassluis, 31 July 1979

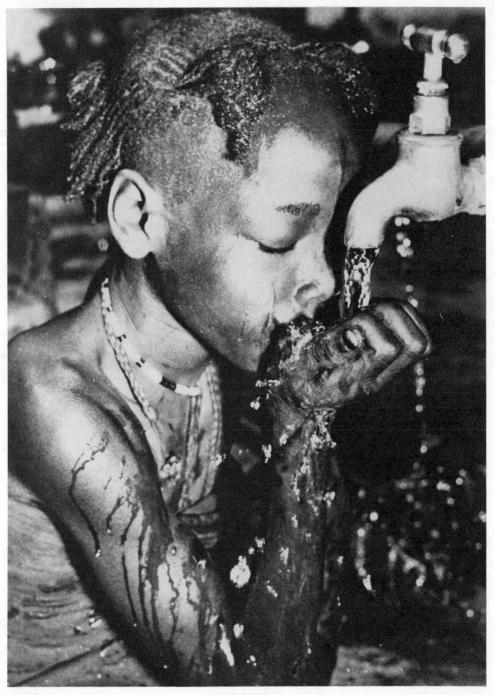

Girl drinking

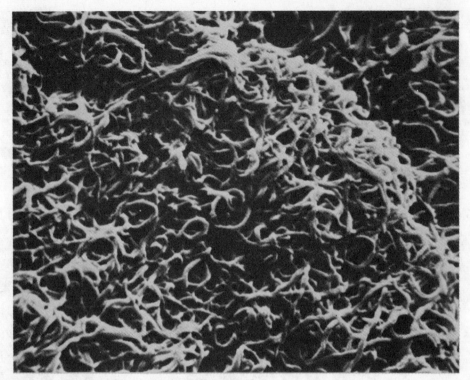
Olfactory epithelium

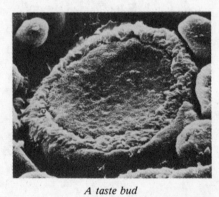
A taste bud

1 Introduction

1.1 General aspects

As a result of the development of powerful analytical techniques, chemists today pay even less attention than before to those quality aspects of water, which are perceptible by the senses. In a more general context Moncrieff (1967) stated that:
"*With the coming of civilization life is less hazardous....As these changes have taken place the senses of smell and taste have lost their sharpness. The process has been slow, but gradually they have dwindled to a shadow of what they were. They did, however, at the zenith of their powers, make such a strong impression on the central nervous system of our ancestors that even to-day their power astonishes us*".
In the past the taste or smell of water, food and air have been considered as important indicators of potential dangers of contamination of these essential factors for human life. There are many examples of the use of the human senses as a warning system for dangers to health. In 1873 Max von Pettenkofer argued against the masking of odours which are according to his view an indication of dangerous air contamination:
"*Eine faule überriechende Luft durch Beimischungen von wohlriechenden Stoffen zu verbessern, ist unthunlich, denn die in einer solchen Luft enthaltenen schädlichen Bestandtteile werden dadurch nicht verändert oder zerstört, es wird nur unser Geruchsorgan in schmeichlerischer Weise betrogen*".
Crompton (1873/1874) even wrote: "*On the use of the sense of smelling in the diagnosis of disease*".
Moncrieff (1967) poses that there is no naturally occurring toxic vapour that is odourless and he relates this statement to Richter (1950) who wrote:
"*Tasteless toxic substances could not have existed widespread in nature in readily available forms at any time in evolutionary history, since in the absence of a taste warning every animal or man that ingested them would have perished. It is more likely that they belong to a group of compounds to which in evolutionary history man and animals have never been extensively exposed*".
Moncrieff further indicates how people used for poisoning purposes particularly those **exceptional compounds which are toxic to most humans at levels which cannot be** sensorily perceived, such as potassium fluoracetate, which has been used by the natives in South Africa for mass poisoning of rival tribes.

As far as drinking water is concerned it is of interest that in the beginning of the 19th century the City Council of Amsterdam regulated the inspection of vessels carrying water from the river Vecht into the city to supply the breweries as well as the population with drinking water of acceptable quality. The regulations included the testing of the water quality in the vessels by a specially appointed water inspector, who tested the purity of the water by tasting it (Leeflang, 1974).

The importance attributed formerly to sensory assessment of water quality also appears from a description by Gaston Tissandier (1873) of those characteristics of water which indicate its safety and wholesomeness as drinking water:
"*Une eau peut être considerée comme saine et de bonne qualité quand elle est fraîche, limpide, sans odeur, quand elle, ne se trouble pas par l'ebulition, quand le résidu qu'elle*

abandonne par l'evaporation est très-faible, quand sa saveur agréable et douce n'est ni fade ni salée, quand elle renferme de l'air en dissolution, quand elle dissout bien le savon sans former de grumeaux, quand enfin elle cuit bien les légumes".

Not only in documents of a century ago but also in the International Standards for Drinking-Water of the World Health Organization of 1971, the sensory aspects of water quality are considered to be of significance:

"Coolness, absence of turbidity, and absence of colour and of any disagreable taste or smell are of the utmost importance in public supplies of drinking water".

However in this type of recent documents the sensory aspects of water quality have been given considerably less attention than the presence of individual chemicals which can be more or less easily detected by analytical instruments and techniques and part of which are known to be of potential danger to health.

On the other hand it is impracticable and physically impossible to measure all chemical compounds in water which are of potential significance to health, as this would lead to the measurement on a routine basis of 50 to 100 or even more water quality parameters. For this reason attention is recently given to chemical indicator compounds and to rapid screening methods for the assessment of the overall effects of water quality on man. As the ultimate goal of all water quality determinations is to indicate potential effects on man, the latter type of measurements is of particular interest for the regular control of water quality.

It is felt that in this respect sensory water quality assessment might play again an important role. Therefore the possibilities as well as the limitations of sensory water quality assessment should be explored in more detail. This is one of the main purposes of this investigation. Furthermore this study is aiming at contributing to a better understanding of the causes and the impact of impaired water quality, as perceived by the chemical senses.

In this introductory chapter a summary of the literature on the nature of sensory water quality assessment and the types of perceptible water constituents will be presented.

1.2 The nature of sensory assessment of water quality

1.2.1 Types of senses for water quality assessment

For assessing the quality of water, organisms possess many different senses. Nerve impulses are generated in specialized receptors after activation by chemical or physical stimuli. In relation to water quality mechanical, thermal, photic, acoustic and chemical stimuli exist.

Man often uses his senses in assessing the quality of water which is intended to be used for boating, swimming, washing, food preparation or drinking.

Water quality will be evaluated first by those human senses which can be stimulated from a distance. Aspects of water quality such as colour and turbidity can be detected from large distances by vision in case lakes or rivers are observed from a high position, like the top of a mountain.

In case an observer is located at the same altitude as the water source smell can be the most powerful sense in detecting quality aspects of water. Examples have been reported where offensive smelling surface water was perceived over a distance of several miles (Maarse and Ten Oever de Brauw, 1972).

Without seeing the surface water, its quality can also be estimated by hearing. The sound of running water can give an impression of its viscosity, while an audible flow-velocity indicates a certain degree of aeration and purification which has led in the past to a preference for running water as a source for community water supply. Illustrative in this respect is the statement of Sir Francis Bacon (1672) that:
"Running waters putrefy not".
Once taken into the mouth water appeals in a multifold way, as sensations such as touch, temperature, chemical irritation, odour and taste are experienced. After assessing the colour, the transparency, the temperature and the smell of water which is intended for drinking, the quality will finally be evaluated in the oral cavity before the water is actually swallowed. In this final process, which includes the quenching of thirst, the chemical senses play an essential role.

Chemical receptors have to be activated by direct contact with chemical compounds. The involved receptors are the olfactory organs, the common chemical or trigeminal receptors and the organs of taste.

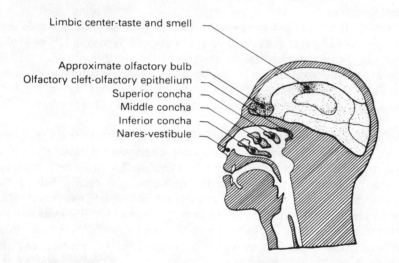

Figure 1.1
The position of the olfactory cleft in the nasal cavity (from Cox. J.P., 1975, Odor control and olfaction, PO Box 175, Lynden, Washington, 98264, 15).

Gustation, olfaction and the common chemical sense are as separate and distinct as vision and audition (Stone and Pangborn, 1968). Of these three senses smell is the most sensitive and irritance the least sensitive sense (Mitchell, 1967).

1.2.2 Human olfaction

1.2.2.1 *Mechanism*

The odour of a substance can only be perceived if the compound is volatile enough to reach the olfactory epithelium in the nose via the air. The olfactory epithelium is located in a narrow passage at the ceiling of the inner nose. This area is somewhat removed from the main respiratory air stream. Only a small part of the inhaled odorant molecules will reach the olfactory epithelium, which fraction can be positively influenced by intensive sniffing. The effective area of the olfactory epithelium, which contains many millions of receptors, is greatly increased by the presence of olfactory hairs which project into the mucous layer and on the surface.

After arrival of the airborne molecules at the olfactory epithelium, transport must take place through the mucous covering the olfactory receptors and a minimum number of molecules must be in contact with the actual receptor cells for perception to take place (Stuiver, 1958) (McNamara and Danker, 1968) (Beets, 1973) (Boelens, 1976).

A number of theories has been proposed for the olfactory transduction mechanism at the plasma membrane of the neurones of the receptor cells.

Firstly Davies and Taylor (1959) developed a puncturing theory, supposing odorous molecules to penetrate the membrane of the receptor cell and thus changing the permeability of the membrane during the short time of adsorption. In this period the sodium-potassium balance of the cell would be disturbed, which initiates the olfactory nerve impulse. According to Wright and Michels (1964) intramolecular vibrations would initiate the nerve impulses.

A steric theory of a lock and key relationship between odorant molecule and receptor site was proposed by Amoore (1952) and Amoore and Venstrom (1966, 1967), however without suggesting a mechanism for the interaction.

At present a plausible explanation for the quality determination mechanism is the occurrence of different types of olfactory receptor proteins on the plasma membranes of the neurones to which odorant molecules can be adsorbed during a short period (Dodd, 1974) (Dutler, 1976) (Menco *et al.*, 1976).

The latter mechanism also complies with factors such as cross-sectional area of the odorant molecule, ability to form hydrogen bonds and electronic polarizability, which were found to account for much of the odorant discrimination (Laffort, 1969). Menevse *et al.* (1977) recently demonstrated the specific involvement of cyclic-AMP in the subsequent transduction step from odorant-membrane interaction into a nerve impulse.

Impulses induced in the olfactory epithelium are subsequently conducted via primary olfactory fibers to the olfactory bulb and via the secondary olfactory pathways to the cerebral hemispheres. According to Moulton (1971) specific receptors and the pattern of odour induced excitation of the different receptor sites, which reaches the brain, may define the perceived quality of the stimulus.

As this survey shows, the mechanism of olfaction is still only partially understood. However for the application of olfaction in the field of drinking water research, several quantitative and qualitative aspects can be effectively described with psychophysical techniques. Psychophysics is the science of the functional relationships between physical stimuli and the sensations of the human observer (Fechner, 1859).

A number of relevant psychophysical techniques will be briefly described in the next paragraphs. For this purpose a recent survey on human psychophysics in olfaction by Köster (1975) has been followed to a large extent.

1.2.2.2 *Odour intensity*

In this section four intensity aspects will be considered:
— the odour detection threshold
— the discrimination threshold
— the relationship between odorant concentration and perceived odour intensity
— the effects of mixing of compounds

Odour detection threshold

The odour detection threshold is rather a statistical than an absolute concept. The chance that a stimulus will be perceived increases gradually over a certain range of intensities. The odour detection threshold of a substance is defined as the concentration at which a subject gives a positive response in 50% of the cases in which the stimulus is presented to him (see figure 1.2).

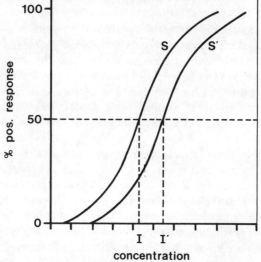

Figure 1.2
Threshold concentration I and I' for the subjects S and S' (after Köster, 1975).

The Odour Threshold Concentration (OTC) of a chemical compound in water can be defined as the concentration at which for 50% of a group of subjects the odour detection threshold is trespassed. In the case of drinking water the odour of the sample will be so weak that most of the consumers can hardly detect it. A valuable theory, which considers different factors which affect the decision process of the observer in the area of close resemblance between the odour of a stimulus and the background odour, is the "signal detection theory" (Swets *et al.*, 1961). This theory considers olfactory signals as appearing on a background of "noise", arising from external events and spontaneous neural discharges (see figure 1.3).

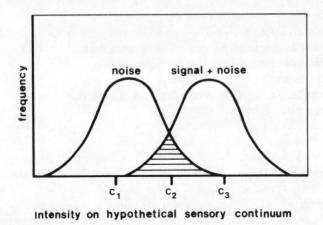

Figure 1.3
Odour signals appearing on a background of noise (after Köster, 1975)

The subject has to decide whether the perceived intensity is caused by the noise or by the stimulus, which decision must become guesswork in the area of overlap. In this area, which represents the actual situation for most of the consumers of drinking water, it is essential which criterion value is used by the subject to decide whether or not the perceived intensity is caused by the stimulus. The decision criterion will depend on the expectancy and motivation of the subject. This actually means that in case smell is expected to be present, it is more often perceived than in the case smell is expected to be absent.

This fact should be incorporated in the set-up and evaluation of field experiments. In the case of laboratory studies it means that factors such as expectation and motivation should be kept as constant as possible during the whole experiment.

For this reason odour threshold measurements should preferably be carried out by means of the so called "forced choice method", in which the subject receives during a presentation a number of stimuli and is instructed to find the stimulus which is different from the others, which are blanks. Even if no difference is observed, a choice has to be made by guessing. In this way five or seven preselected stimuli of different concentrations are presented together with the blanks a large number of times to the subjects in a randomized order.

Individuals differ appreciably in sensitivity to odours. Inter-individual differences in sensitivity show a range of 100 - 10.000 as indicated by threshold determinations (Baker, 1963) (Brown *et al.*, 1968) (Zoeteman and Piet, 1973). Furthermore the person most sensitive to one compound may be least sensitive to another. Small parts of the population are even anosmic to certain compounds (Amoore *et al.* 1968) (Harper, 1972). Koelega and Köster (1974) found that adult females are more sensitive to some odorants than adult males. With increasing age, particularly from the age of 60 years onwards, the sense of smell shows a progessive reduction in sensitivity (Hughes, 1969). Each individual can also show large variations in sensitivity in time due to factors such as adaptation, fatigue, motivation and environmental conditions.

Adaptation is a reduction of the sensitivity of a sense organ as a result of stimulation, from which the sensory system recovers gradually after cessation of the stimulation. Adaptation does occur also when very short stimuli of near threshold intensity are used. According to Köster (1971) it may take 60 seconds until complete recovery is reached after such weak stimuli, which finding is of direct importance to the set-up of panel experiments with samples of drinking water.

Odour detection thresholds furthermore vary considerably from one substance to the other. Among the more than 10.000 odorous chemicals, compounds with high OTC values in water as ethanol, differ a factor 10^8-10^9 with compounds with low OTC values such as mercaptans and β-ionone (Theranishi, 1971). Only the latter type of compounds can be of importance to the flavour of drinking water, due to the low concentrations of organic compounds, present in water destined for human consumption.

Discrimination threshold

The discrimination threshold, often called the just-noticeable-difference (jnd) is defined as the difference between odorant concentrations, above the detection threshold, which is just detected by the subject in 50% of the cases. Fundamentally the detection and discrimination problem can be treated in the same way. Usually a paired comparison method is applied for measurements of the jnd for a compound.

According to Weber's law the concentration change necessary to produce a just noticeable difference in sensation is proportional to the stimulus intensity. Until recently it was accepted that two concentrations of the same odorant cannot be distinguished if these concentrations differ less than 15-30% (Kniebes *et al.*, 1969). Cain (1977) however showed instances in which subjects resolved differences in concentration of odorants of only 5%. So it may be expected that even such fine variations in drinking water quality can be noticed by the consumers.

Odorant concentration and perceived odour intensity

Olfactory sensation grows as a non-linear function of odorant concentration (Engen, 1971) (Cain and Moscowitz, 1974). Fechner (1859) proposed a logarithmic relationship between sensation (S) and odorant intensity (I) : $S = k \log I$, in which k is a constant. Stevens (1957) however has proposed a power law, reading: $S = I^n$, in which n is a constant.

The rate of growth of odour intensity at increasing concentrations of the odorant can vary considerably from one compound to the other (Henion, 1971) (Dravnieks, 1972). At high odorant concentrations the perceived odour intensity reaches a point of saturation. Several scaling methods have been developed to describe the perceived odour intensity. In category scaling successive numbers are assigned to descriptive terms, for instance ranging from *"very weak"* till *"very strong"* and for each stimulus the average value of the scale numbers corresponding with the obtained category judgements is calculated. According to the method of magnitude estimation a subject has to assign numbers to each of a series of intensities, presented in a random order, in such a way that he feels that the ratios between these numbers correspond with the ratios between the observed intensities. Any number may be used in this method. A third method, cross-modality matching, consists of expressing the strength of the sensation of a subject by matching it with an equally strong sensation in another sense modality, such as audition.

In practice an estimation of the odour intensity of water or air is often made by diluting the sample until the odour detection threshold for the present mixture of compounds is reached. The necessary number of dilutions, called the Odour Number (ON), is considered as a measure for the odour intensity. From a theoretical point of view this concept is wrong, as the concentration - odour intensity relationship is different for different compounds. This method, although often used for raw water types, is of no value for drinking water of which moreover the odour intensity will be below the detection threshold for a large part of the population. In this case a more direct method than the dilution method, such as category scaling, is needed.

Effects of mixing

As drinking water generally contains a large variety of substances which can be perceived by the chemical senses the effect of the presence of mixtures of perceptible compounds should be considered. The combined olfactory effect of the presence of the individual compounds can be a summarization of the individual intensities of the odorants (additivity), or enhancement of the odour intensity (synergy) or can be a decrease of the resulting olfactory intensity (compensation) (Zwaardemaker, 1907). According to Köster (1969) synergy is very rare, additivity is more likely to occur in the case of equal mixing ratios of odorants, while compensation occurs frequently.

Final consideration

Generally it must be stated that due to the large interindividual and intraindividual differences in sensitivity to odours, groups of sufficient size are needed in olfactory water quality assessment in order to obtain reliable and reproducible judgements.

This need for large groups, although a practical disadvantage of sensory evaluation of water quality, should be overcome by mobilizing larger groups of subjects for regular sensory assessment of drinking water quality.

1.2.2.3 *Odour quality*

Experienced perfumers can distinguish effectively about 100 different odour qualities (Harper, 1972). In general, compounds with similar odour may belong to several different chemical classes and compounds of the same chemical class may smell completely different. Mixing of odorous compounds can not only result in effects such as masking but a new odour character may emerge. Odour character can also change with the concentration of the odorant.

Odours have been classified on the basis of similarity and dissimilarity in

perceptual quality (Zwaardemaker, 1895) (Henning, 1916), (Crocker and Henderson, 1927) (Yoshida, 1964). Woskow (1968) found that similarity rating probably is strongly related to the hedonic dimension of the odour character of compounds. Acceptability or hedonic attribute of an odour is strongly dependant on the context (Dravnieks, 1972). It is generally assumed that the large interindividual variation in judgement of the pleasantness of odours mainly is based on associations relating to individual experiences. Studies by Stein et al. (1958), Moncrieff (1966) and Engen (1974) have indicated that children tend to be more tolerant to unpleasant odours than adults. Engen (1974) states in this respect:
"The older the children the more the preference agrees with those of adults and (there is) evidence to the hypothesis that discrimination of hedonic attributes of odours is learned. There may be no inherently unpleasant odours".
Furthermore there is evidence that there is a relation between familiarity and tolerance or liking of odours (Foster, 1963). If an odour is unfamiliar to a person, he will tend to reject it. For a specific product, such as drinking water, a set of typical odour quality aspects, an odour language, has to be developed. A suitable method for measuring subjective odour quality judgements consists of the use of similarity rating scales for different odour quality aspects, while each quality aspect is defined by a standard solution of one or more chemical compounds. Other measuring techniques for similarity between odours are ordinal scaling methods (Coombs, 1964) or the use of a confusion matrix (Köster, 1975).

1.2.3 Human gustation

1.2.3.1 *Mechanism*

The receptors for taste are the taste cells. These are concentrated mostly in the taste buds, which are composed of a group of taste cells in a cluster with supporting cells. On the tip of each cell are cilia which extend into a pore which opens to the mouth (McNamara and Danker, 1968). There are about 3.000 - 10.000 taste buds in the mouth, most of which are located on the upper surface of the tongue, at its tip, sides and rear surfaces. There is an area in the middle of the tongue which has no taste buds. Taste cells constantly degenerate and regenerate and have a life-time of a few days (Beidler, 1964). The total number of taste buds slowly decreases with increasing age.

In gustation, the stimulant molecules dissolved in saliva enter the pore of the taste bud and a contact between this aqueous solution and the receptor sites is established. The situation in the polymolecular layer of saliva in immediate contact with the sites at the cell surface is virtually the same as in olfaction, and the same model of the interaction scene can be used for both senses (Beets, 1973) (Hansen et al., 1976) (Kijima, 1976).

According to recent reviews of Meiselman (1972, 1976) there is no evidence that specific taste nerves exist for the four classical taste qualities: sour, salty, bitter and sweet. In place of specificity of taste receptors it is suggested that some form of neural patterning, which is containing the information about taste quality is decoded at the higher brain centers.

1.2.3.2 *Taste intensity*

As in the case of odour intensity, taste intensity is influenced by the stimulus concentration. Data of Moskowitz (1970a, 1971, 1973) suggest that the perceived taste intensity is a power function of the stimulus concentration.

Taste intensity is also affected by factors like viscosity of the liquid phase (Moskowitz, 1970b) and the fact that drinking of water from a glass is an intermittent process. Meiselman and Halpern (1973) reported the occurrence of enhancement of salt taste intensity during pulsatile stimulation of one second duration. Several authors reported on the effect of water temperature on taste intensity. According to some, optimal taste sensitivity is obtained at 22°C (Griffin, 1966) (Fisher, 1971) (Pangborn and Bertolero, 1972). Others could not demonstrate an effect of temperature on the perceived taste intensity of certain compounds (Stone *et al.*, 1969) (Moskowitz, 1973). An increase in sensation can also be temporarily obtained by tongue movements which bring previously unexposed receptors in contact with the stimulus.

Furthermore taste research and particularly studies on water taste have to take into consideration the constant presence of an adapting condition within the mouth. Pure water is not tasteless according to a study of Bartoshuk (1974) and drinking water will only be tasteless when it contains salt ions in concentrations which are not too much different from the concentrations in the saliva.

Cross adaptation was found for substances representing each of the traditional taste quality categories (McBurney, 1969). Bitter taste can be quite lasting, which has been related to the affinity of bitter tasting compounds to the skin. According to McNamara and Danker (1968) bitter taste may last for over 1 minute even after rinsing.

The sense of taste is much less sensitive than the sense of smell. Bitter tasting compounds, which generally show the lowest taste threshold values, have thresholds at 10^6 times higher levels than the most odour intensive compounds (Moncrieff, 1967). Besides bitter tasting organic compounds, such as quinine, ions from metals such as iron and copper can be tasted at concentrations of 1-10 mg/l (Cohen *et al.*, 1960).

Acids such as tartaric acid and hydrochloric acid can be perceived at concentrations of approximately 50-100 mg/l while most salts have taste thresholds between concentrations of 100-1000 mg/l (Lockhart *et al.*, 1955) (Bruvold and Pangborn, 1966) (Moncrieff, 1967). The taste threshold for sweet tasting sugars such as sucrose lays between 1-10 g/l (Moncrieff, 1967). Within mixtures of compounds representing different taste qualities, the tastes usually suppress each other (Meiselman, 1976).

Individual differences in taste sensitivity can amount up to a factor 100-1000 (Geldard, 1953) (Fisher, 1971). Several substances are known to which part of the population is relatively insensitive like phenylthiocarbamide. Fisher (1971) showed that age does not significantly affect taste sensitivity to 6-n-propylthiouracil and quinine. However deterioration of taste sensitivity for bitter tasting compounds has been related with age for heavy smokers (Kaplan *et al.*, 1964) (Fisher, 1971). According to Soltan and Bracken (1958) a greater number of females compared to males taste quinine as bitter. Kroeze (1971) found about two times lower threshold taste concentrations in water for quinine sulphate and sodium chloride for females compared to males. Kaplan and Fisher (1965) showed that bitter tasting compounds like quinine were perceived with greater sensitivity by the majority of females during the menstrual period, which effect could however not be clearly demonstrated by Kahn (1965). Such individual differences have to be considered in selecting and evaluating the results of panels.

1.2.3.3 *Taste quality*

The four major descriptive terms for taste quality are saltness, sourness, sweetness and bitterness, although there is no substantial electrophysiological evidence for the actual existence of four basic tastes (Meiselman, 1972).

Hydrogen ions are the activator of the sour taste. Sodium chloride is generally used as a standard for a salty taste. Other inorganic compounds can taste bitter or sweet (Moncrieff, 1967). Many different compounds from sugars to lead salts have a sweet taste. Bitter tasting substances are caffeine, nicotine and several salts of magnesium and iodine. Furthermore salts can have several tastes. Magnesium sulphate is bitter at the back and salty near the front of the tongue.

There are several substances, which have taste modifying properties (Bartoshuk *et al.*, 1969). The best known among these are gemnemic acid, derived from *Gymnema sylvestre* which strongly depresses the sweetness of sweet-tasting materials, and miraculin, derived from the berries of *"miracle fruit"*, *Synsepalum dulcifirum*, which sweetens acidic materials. Aspects such as modification, masking and hedonic attribute have not yet been extensively studied in relation to the taste of drinking water.

1.2.4 Some effects related to chemoreception

According to Stone and Pangborn (1968) stimulation of one sense organ influences to some degree the sensitivity of the organs of another sense. Many sensations commonly attributed to taste are in fact a combination of taste and odour, or even only due to odour. Besides sensory interaction there exists a relation between chemoreception and drug reactivity. Fisher (1971) states that a sensitive taster will generally also need a lower dose of a drug to elicit a specific pharmacological effect. He has illustrated this by comparing the taste threshold for the drug trifluoperazine and the applied dose to a group of chronic schizophrenics (Fisher *et al.*, 1965) (Knapp *et al.*, 1966). A confirmation of the proposed relationship has been given by Joyce *et al.* (1968) who showed the relation between taste sensitivity to the drug hyoscine butylbromide and amongst others heart rate in healthy medical students.

This relationship would imply that the use of chemoreception as a warning mechanism could be adapted to a certain extent to the specific needs of the individual involved.

1.3 Perceptible substances in drinking water

1.3.1 Historical developments in The Netherlands

Since the first waterworks of The Netherlands was established in 1854, near the city of Amsterdam (Leeflang, 1974), ground water has been the major source of potable water supply in The Netherlands. Substances in ground water, such as iron, humic acids and hydrogen sulphide, could generally be sufficiently removed by aeration and sand filtration and were of course considered of less importance to health than the bacteriological characteristics of the water. At the beginning of this century the analysis of drinking water however was still based to a large extent on quality aspects detectable by the senses. Generally a data sheet started by describing the colour, odour, taste and transparency of the water followed by noting the absence or the presence of a trace of ammonia, iron etc. (Hoogdrukwaterleiding van de Gemeente Zwolle, 1908).

Those waterworks on the other hand which used surface water as a source were confronted with severe problems during the last decades due to the presence of increasing quantities of contaminants.

Very illustrative in this respect is the history of the waterworks of Rotterdam. This city has been supplied with drinking water derived from river water since the beginning of the supply in 1874. In de 19th century a simple sand filtration was sufficient to produce wholesome drinking water from water of the river Rhine.

In 1921 the river Rhine showed relative low flows and the yearly report of the Rotterdam waterworks records that the drinking water had a very disagreeable taste during the last months of the year (Drinkwaterleiding Rotterdam, 1921). In 1928 serious trouble developed due to a *"musty"* taste of the water in the winter periods.

Also in other cities like Dordrecht, the bad taste and odour of the water of the Rhine was noticed. Particularly Heymann (1931/1932) reported on the *"earthy"* taste of the water. His daily observations of the taste of the Rhine water at Rhenen in the period 1928-1932 indicate that most of the adverse taste of the Rhine water must have been caused by waste water discharges of coke furnaces in the drainage area of the river Emscher in Germany, one of the tributaries of the river Rhine. Although chlorinated phenols have been postulated as the cause of these problems, this has never been proven. The serious organoleptic problems with the Rhine water in The Netherlands even resulted in the installation of a Committee on Taste and Odour of River Water, by the president of the Health Council on June 7, 1929, which committee however never prepared a final report during the 5 years of its existence (Leeflang, 1974).

The treatment facilities in Rotterdam were extended in 1931 with rapid sand filters before the slow sand filtration step. In December 1933, dosing of activated carbon powder started. The dose amounted to 7 mg/l in the winter period in 1934.

After the second world war the period in the summer, during which no activated carbon dosing was needed, became shorter and shorter and in 1954 the carbon dosing had to be applied the whole year. In 1958 the maximum dose amounted to 27 mg/l and in 1959 it was no longer possible to remove the taste completely by carbon treatment, as the dose had to be limited to a maximum of 30-40 mg/l to maintain practical operation conditions for the filters. The situation improved considerably only in 1973 after the city started to use water of the river Meuse instead of the river Rhine.

During the 20th century similar problems to those encountered by the Rotterdam water works emerged at those supplies which used water from the river Rhine by means of bankfiltration or after storage in the dunes or in open reservoirs. It is anticipated (Structuurschema - 1972, 1975) that surface water will be used more and more because the national ground water resources are insufficient to meet the expected growing water demand. In 1971 surface water contributed up till 35% of the total water demand of 1700 million m^3/year in The Netherlands, which figure might become 53% in the year 2000 in case the total water demand increases to a quantity of 4000 million m^3/year.

These expected developments emphasize the need to consider in more detail the causes of perceptible contamination of drinking water.

In the following section substances causing turbidity, colour, odour and taste of water will be discussed. As taste and odour compounds seem to be most indicative for unacceptable water contamination, attention will particularly be given to these two types of perceptible water contaminants.

1.3.2 Substances causing turbidity of water

The turbidity of water is due to suspended and colloidal matter, the effect of which is light scattering and diminished light penetration. Turbidity can be caused by micro-organisms or organic detritus, silica, clay or silt, fibres and other materials (McKee and Wolf, 1963). In general aerobic ground water has a low turbidity while turbid river water contains 10-100 mg/l or more of suspended matter. In drinking water mineral substances like zinc, iron and manganese compounds can cause turbidity. High concentrations of 30 mg/l or more of zinc give water a milky appearance (Kehoe *et al.*, 1944) and above levels of 5 mg/l zinc causes a greasy film on boiling (Howard 1923).

Removal of water turbity is of importance as it guarantees to a certain degree the absence of many kinds of materials adsorbed on suspended matter. Furthermore water with a good transparency can be more effectively disinfected by oxidants. Water can show a higher turbidity at the tap than at the treatment plant due to pick-up of sediments of iron and manganese hydroxides or due to corrosion of copper, zinc or iron containing piping (Zoeteman and Haring, 1976).

1.3.3 Substances causing colour of water

Drinking water derived from ground water can contain high quantities of fulvic and humic acids of natural vegetable origin, giving the water a yellow or brownish colour. This type of organic substances can chelate metal ions, thereby interfering with coagulation (Hall and Packham, 1965) or rendering iron and manganese more difficult to remove during treatment (Shapiro, 1964) (Zoeteman, 1970a). Drinking water can be coloured due to the presence of small algae that have passed through the filters of a plant, treating water from a lake or a reservoir. Water colour can also be caused by industrial compounds (Meijers, 1970).

Furthermore metallic compounds like copper, iron and manganese can cause colour problems. Slightly blue or green colouration appears in water transported through copper pipes when at least 5 mg/l of insoluble copper corrosion products is present in the water (Page, 1973). Dissolved iron (III) chloride imparts a brownish-yellow colour. Iron and manganese are also objectionable in drinking water because of spotting of laundered clothes. At levels below 0.3 mg/l of iron and 0.05 mg/l of manganese the red iron stains (Buswell, 1928) and black manganese stains (Griffin, 1960) do not manifest themselves.

1.3.4 Compounds causing odour of water

1.3.4.1 *General aspects*

A large variety of odorous compounds have been found in drinking water. These substances can be present in the raw water source, can be formed during treatment, particularly by chlorination, or can be introduced into the water during distribution (Drost and Zoeteman, 1976). With a few exceptions, the odours in drinking water are caused by organic compounds. The types of odorous materials related to the three causes mentioned above will be considered in some more detail below.

1.3.4.2 *Odorous compounds originating from the raw water*

Ground waters are generally free of odours after sufficient aeration to remove hydrogen sulphide. The odorous substances in water of rivers and lakes can be much more persistent. These substances can be introduced into the surface water by municipal and industrial waste water discharges or can be formed by organisms (Rosen *et al.*, 1963) (Bays *et al.*, 1970) (Kölle *et al.*, 1970).

A survey of four surface waters in The Netherlands in 1971 showed a significant increase in the number of odorous organic compounds during the summer (Zoeteman and Piet, 1972/1973) and a rapidly changing pattern of the type of compounds over the year. It has been indicated that in river water such as water of the river Rhine with a high load of industrial odorous substances, the seasonal changes in water odour are determined by evaporation and biological mineralization processes (Zoeteman, 1970b) (Oskam and Rook, 1970) (Zoeteman and Piet, 1974a). In case of more or less stagnant eutrophic surface waters the influence of biologically produced substances, such as geosmin and 2-methylisoborneol, can become dominant temporarily as has been shown for the river Meuse in 1972-1973 (Zoeteman and Piet, 1974a) (Zoeteman and Piet, 1974b). Industrial compounds which have been shown to be of potential significance for the odour of drinking water are particularly o-chlorophenol, o/p-dichlorobenzene, indene, 2-methylthiobenzothiazole, naphthalene and 1,3,5-trimethylbenzene (Zoeteman *et al.*, 1975).

In general all relatively stable compounds with threshold odour concentrations in water of less than 1 microgram/litre are of potential interest as most organic compounds in drinking water are not present in concentrations above this level.

1.3.4.3 *Odour causing compounds introduced during water treatment*

During the processing of water from the water intake to the point where it enters the distribution system the role of metabolites or decaying products of organisms and reaction products of chemical oxidants have to be considered. Particularly in storage reservoirs the earthy smelling geosmin and 2-methylisoborneol can be produced by certain blue-green algae and streptomycetes (Safferman *et al.*, 1967) (Gerber, 1968) (Medsker *et al.*, 1968) (Leventer and Eren, 1969) (Rosen *et al.*, 1970) (Piet *et al.*, 1972) (Silvey *et al.*, 1972) (Vlugt *et al.*, 1973).

After massive dying of algae the water can also contain organic sulphur compounds like dimethyldisulphide. In the case of bank filtration of river water it has been reported that certain sesquiterpenes might be formed (Koppe, 1965). The possible formation of intermediate degradation products such as aldehydes and fatty acids of branched aliphatic hydrocarbons during bank filtration of Rhine water has been studied. These studies did not indicate a considerable role of this possible mechanism of odour production, although 2-methylbutyric acid was tentatively identified as a product present in the bank filtered water (Boorsma et al., 1969) (Zoeteman et al., 1971).

Chemical oxidation by chlorine treatment has been known for a long time as a possible cause of offensive odour in drinking water. Particularly chlorination of water containing phenol has been reported to increase the odour of the water due to formation of chlorinated phenols, which have an OTC at roughly 1000 times lower levels than phenol itself (Burtschell, 1959).

However, also other chlorinated compounds like the chlorinated anilines and benzenes show much lower odour thresholds than the unchlorinated compounds (Zoeteman and Piet, 1974a), although it is uncertain at present whether or not such chlorinated compounds are formed in practice. Among the haloforms which are formed during water chlorination from humic and fulvic acids (Rook, 1974), chloroform can be perceived sensorily (Zoeteman et al., 1975) at concentrations above 10-100 $\mu g/l$, although its sweetish smell is not as adverse as that of chlorinated phenols. In practice the overall effect of chlorination generally is a slight improvement of water taste and odour as long as chlorine itself is not present in perceivable quantities.

1.3.4.4 *Odour causing compounds introduced during distribution*

Odours can sometimes be introduced during distribution as a result of the release of organic compounds from protective coatings like bituminous materials, which can release aromatic hydrocarbons like naphthalene. The most frequently occuring cause of odours introduced during distribution is growth of micro-organisms which form earthy or musty smelling metabolites like geosmin. This occurs particularly in waters, containing relatively high quantities of biologically degradable organic substances, and at relatively high temperatures. The latter can be the case in long supply pipes in large blocks of flats or during the summer in case the water is derived from surface water (Packham, 1968) (Zoeteman and Haring, 1976)

1.3.5 **Mineral substances causing taste of water**

Substances in drinking water which are perceived by the sense of taste are generally inorganic compounds which are present in much higher concentrations in drinking water than organic micropollutants.

It has been generally established that among the salts present in drinking water the sulphates and hydrocarbonates are less affecting taste intensity and taste quality than the chlorides and carbonates (Bruvold and Pangborn, 1966) (Pangborn et al., 1971).

The taste threshold concentration of sodium carbonate (about 75 mg/l) is about 15 times lower than the threshold level of sodium hydrocarbonate (1060 mg/l) (Lockhart et al., 1955). Chlorine is reported to be detectable, depending on water pH from 0.075

mg/l at a pH of 5.0 to 0.450 mg/l for a pH of 9.0 (Bryan et al., 1973), which indicates that chlorine will be tasted by the consumer in many cases where chorine disinfection is practiced.

Among the metal ions which can be present in drinking water, iron could be tasted in distilled water by the most sensitive 5% of a panel at 0.05 mg/l, copper at 2.5 mg/l, manganese at 3.5 mg/l and zinc at about 5 mg/l (Cohen et al., 1960). Particularly iron might affect water taste in practice.

1.4 Outline of the work

As stated in section 1.1 the main purpose of this investigation is the exploration of the possibilities and limitations of sensory water quality assessment as a rapid overall indication of the quality of water and of potential effects on the health of consumers. Such an exploration should lead to identification of those compounds generally responsible for objectionable water taste and odour, in order to avoid their introduction into the water at the source. The latter could contribute to including individual compounds in the list of "substances having a deleterious effect on taste and odour of products derived from the water for human consumption", which list is part of the directive of the Council of the European Communities (1976), relating to the contamination caused by certain dangerous substances which are discharged into the aquatic environment of the Community. The same list is also part of the "Convention against the chemical pollution of the River Rhine", signed at December 3, 1976 by the ministers concerned.

In order to quantify problems relating to taste and odour of drinking water in The Netherlands, an inquiry was carried out among a representative part of the Dutch population. The design and results of this inquiry are decribed in Chapter 2.

For a selection of 20 communities in The Netherlands tapwater samples were presented to a national panel selected for objective assessment of taste and odour aspects. The results of the water quality assessment by the panel and recommendations, relating to a methodology of rating the perceptible water quality, are presented in Chapter 3.

The same water samples, used for taste and odour assessment by the panel, were analysed for the presence of minerals and in particular organic micropollutants which could be responsible for impaired water taste and odour. The results are discussed, in relation to the taste and odour assessment by the panel, in the Chapters 4 and 5.

Chapter 6 deals with the sensory assessment of water quality in relation to health protection aspects.

Finally, Chapter 7 gives a summary of practical methods for sensory water quality assessment and of possible measures to reduce taste and odour problems against the background of the need to use raw waters for potable water supply which become increasingly loaded with more or less treated streams of waste water. The need for further studies and a recommended approach in advanced sensory water quality measurement is discussed.

Atomic absorption spectrofotometer

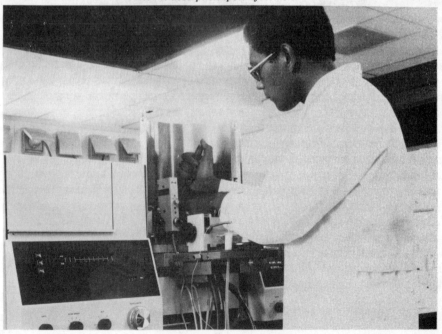
Injection of a water sample in a flameless atomic absorption spectrofotometer.

Water storage reservoir at the Biesbosch for the water supply of Rotterdam

2 Sensory assessment of drinking water quality by the population of The Netherlands

2.1 Introduction

According to different studies, the taste of drinking water is an important factor in the overall assessment of water quality by the consumer.

A survey of consumer complaints in the London Metropolitan Area over the period 1957-1964 indicates that 44.5% of the complaints were related to the taste and smell of the water, 15.3% to rust particles or sediment and 9.6% to discoloured water (Windle Taylor, 1965).

A Gallup Poll (1973) carried out among 3000 adults in the USA showed that the most frequent type of complaint concerned a bad taste of the water (14.1% of total responses), followed by complaints on discoloured or dirty water (8.9%) and bad smell (5.7%). An inquiry in the Federal Republic of Germany held in October 1975 among 2000 persons by the Institut für Demoskopie Allensbach (Anon., 1977) showed that 22% of the national population rated the taste of their drinking water as unpleasant.

An explorative study among 1000 visitors of an exhibition in Amsterdam in September 1972 indicated a strong relation between the type of raw water used for potable supply and water taste assessment. Furthermore impaired water taste seemed to result in a reduction of the actually consumed quantity of water as drinking water (Zoeteman and Piet, 1973).

A general inquiry carried out in February 1974 on behalf of the Dutch Waterworks (VEWIN) among 2500 adults in The Netherlands (Buyzer, 1974) showed that 7% of the population rated their tapwater quality as bad, while within this category 36% indicated a bad taste as criterion for this judgement.

As these studies do not relate sensory quality assessment to possible causes of reduced quality or to the impact on consumer behaviour, it was of interest to carry out an inquiry particularly aiming at these types of questions. For this purpose questions relating to sensory assessment as well as general appraisal of tapwater quality were formulated. Also several questions were posed which could give an indication of factors influencing the individual judgements. This set of questions on sensory water quality assessment was added to an already planned questionnaire, set up in co-operation with the Central Bureau for Statistics (CBS), Voorburg, The Netherlands. The planned questionnaire concerned water consumption habits in households in relation to human exposure to metals via tapwater due to release of metals from piping materials. By this combination the interpretation of differences in water consumption patterns could be improved and on the other hand the effect of sensory assessment on water consumption could be evaluated.

In this chapter the results obtained on sensory water quality assessment and on general water quality appraisal will be discussed while in Chapter 6 the effects of these factors on water consumption will be considered. First a description of the inquiry will be given, followed by a presentation of the general results. Furthermore possible general causes of impaired perceptible quality of drinking water are discussed.

2.2 The inquiry

2.2.1 The questionnaire

The total questionnaire consisted of 4 parts:
1. General questions relating to the members of the household, including sex and age of the individuals, educational standard of the head of the household and questions on the type of house (one family house, bungalow, flat, etc.) age of the house and type of material used for the domestic plumbing.
2. Questions relating to water consumption habits of each member of the household. Individuals had to indicate the amount of water consumed at their homes or elsewhere during two days. One day was reckoned to be from 07.00 - 07.00 hours. Units were defined as follows:
   - gulp of water : 30 ml
   - cup (for water, tea or coffee): 125 ml
   - beaker or glass, small size : 150 ml
     (for water or lemonade)
   - beaker or glass, large size : 250 ml
     (for water or lemonade)
   - cup or plate (for soup) : 250 ml
   - baby food flask : ml's per flask to be stated

   The average volumes indicated, resulted from testing of several types of relevant materials from 10 households. These data were used to calculate the daily consumed total amount of water per person as water, tea, coffee, soup or lemonade.
3. Questions on general consumption habits in the households, such as flushing the tap before use, use of hot water for tea or coffee etc..
4. Questions on sensory water quality assessment for members of the households.

The questionnaire relating to sensory water quality assessment consists of questions on colour, temperature, odour and taste of the water, which had to be answered by indicating on a 3- or 5-point category scale the item which was nearest to the personal judgement. Furthermore the quality of the taste and smell could be indicated by crossing one out of five qualifications. Also the taste of the home brewed tea had to be described on a 3-point category scale as well as the occurrence and eventual offensiveness of a skin on the tea if a cup of tea is poured. The acceptability and safety of the tapwater from a health point of view had to be indicated on a 5-point respectively 3-point category scale. Finally several questions were included which could possibly explain certain attitudes in sensory assessment of water quality, such as the judgement on taste of tapwater distributed elsewhere, the opinion on the type of source of the tapwater and the opinion on air quality at the locality and on air quality at the place of work.

2.2.2 Preparation and realization

Before the inquiry was carried out, a pilot inquiry was composed in co-operation with sociologists and other specialists of the CBS and was tested in February 1976 among 110 households to trace any shortcomings or practical problems. Some indicative results of this pilot study have been presented in a joint paper with the CBS (Zoeteman *et al.*, 1976). After improving the phrasing of several questions the final inquiry was held in June 1976.

In order to obtain a representative picture, relating to the whole population of The Netherlands, a sample of the file of addresses of households was taken, which was stratified according to water hardness and water source. Based on the resulting distribution of individuals, according to criteria such as age, sex and educational standard of the head of the household, the representativity of the sample could be evaluated. A total number of 2000 addresses of households in The Netherlands were selected by the CBS.

After a postal announcement, the households were visited personally by instructed collaborators of the CBS, to introduce the questionnaire and afterwards to go through the completed questionnaire, together with the individuals concerned to clarify any remaining problems.

A response was obtained from 4630 individuals belonging to 1472 households located in 89 communities. Within the total study the part on sensory assessment concerned a total number of 3073 individuals of 18 years and over.

2.2.3 Characteristics of the sample of the population

As far as the type of house is concerned the sample of the population rather closely resembled the national situation, as indicated by table 2.1.

Table 2.1: Estimated distribution of types of houses in The Netherlands in 1976 and distribution within the sample (n = 1472)

Type of house	National estimate	Sample
One family house, bungalow, mansion	68	66
Flat	31	28
Others	1	6

The national situation has been estimated by the CBS on basis of the census of 1971 and known developments since that year. Among the households of the sample 40% were constructed before the year 1950 while the CBS estimated that this figure is 41% for the national situation. Table 2.2 gives a survey of the categories within the sample of the population relating to areas with tapwater derived from different sources and with tapwater belonging to different hardness categories.

The data relating to the national situation are based on a survey of the VEWIN (1976). Within the sample of the population the households supplied with bank filtered water were slightly overrepresented and the households with a water hardness between 2-4 meq/l were rather poorly represented.

Table 2.2: Distribution of water hardness and type of source of drinking water within the sample of the population (n = 4620) and the population of The Netherlands*

Hardness category (milli-equivalent/ litre, total hardness)	Type of water source (% of total)			
	Ground water	Bankfiltered surface water	Surface water	Total
0 - 2	15.0 (12.8)	- (-)	- (-)	15.0 (12.8)
2 - 4	23.5 (34.0)	- (0.2)	7.1 (9.9)	30.6 (44.1)
4 - 6	19.9 (16.8)	3.9 (2.1)	12.6 (8.5)	36.4 (27.4)
> 6	7.8 (5.9)	1.6 (0.3)	8.6 (9.5)	18.0 (15.7)
Total	66.2 (69.5)	5.5 (2.6)	28.3 (27.9)	100.0 (100.0)

*Data in brackets apply to the national population

A comparison of the age distribution within the sample and the national population in 1976 is given in figure 2.1. Although some slight deviations exist, the age distribution of the individuals of the sample fits that within the country quite well.

Some characteristics of the group of individuals of 18 years and over which were involved in the sensory assessment of tapwater quality, and which belonged to the described sample of the population, are summarized in table 2.3.

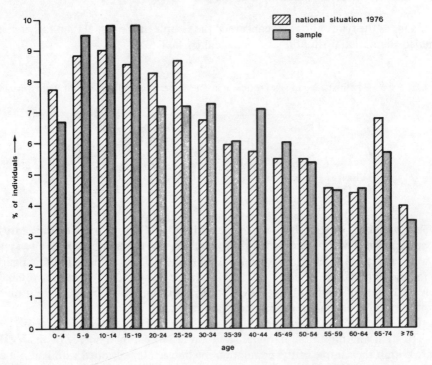

Figure 2.1
Distribution of age of the sample of the population and the national population of the Netherlands in 1976.

Table 2.3: Educational standard of the head of the household, type of house, sex and age distribution within the sample of the population (n = 3073)

Age category (years)	% of total per column										
	Education head of household			Type of house					Sex		Total
					Flat						
	Primary education	Secondary education	Further education	One family house or comparable	1st/2nd floor	3rd/4th floor	5th and higher floor	Other	Male	Female	
18 - 25	13.8	15.3	9.8	14.9	11.6	13.9	7.9	16.3	16.3	14.7	15.5
25 - 35	11.6	30.1	29.3	22.5	20.7	32.1	19.8	17.4	22.0	21.9	22.0
35 - 50	23.1	32.5	30.3	30.1	22.4	20.6	23.8	34.7	28.1	28.0	28.0
50 - 65	27.1	16.5	19.2	20.7	23.7	18.2	20.6	24.0	21.0	20.7	20.8
65 and over	22.4	5.5	11.5	1.8	22.2	14.6	28.6	7.7	12.6	14.6	13.7
n	1286	1500	287	2121	465	165	126	196	1521	1552	3073
% of total per question	41.9	48.8	9.3	69.0	15.1	5.4	4.1	6.4	49.5	50.5	100.0

2.3 Sensory assessment of tapwater quality and possible causes of quality impairment

2.3.1 Visual aspects

As indicated by table 2.4 drinking water is colourless and clear for 84% of the population, while only 0.3% finds it always turbid or coloured.

Table 2.4: Visual assessment of drinking water and type of raw water source (n = 3073)

Type of raw water source	Colour and turbidity assessment % of total scores per raw water category					Contains sometimes brown or black particles
	Always clear and without colour	Sometimes slightly coloured	Mostly a colour which is not offensive	Usually offensive colour or turbid	Always coloured or turbid	
Ground water	87.0	11.0	1.1	0.6	0.3	6.6
Bankfiltrate	73.7	18.9	4.0	3.4	0.0	9.7
Stored surface water	79.0	15.4	3.7	1.5	0.4	8.5
Total (% of total)	83.9	12.8	2.0	1.0	0.3	7.3

23

Generally the requirements of being clear and colourless as well as the need for the absence of brown or black particles are best met by the drinking water derived from ground water and least by the bank filtered type of drinking water. The higher incidence of the presence of brown and black particles in the tapwater of surface water supplies suggests that the presence of iron and manganese compounds, of which such particles consist might contribute to the colour and turbidity of some types of drinking water derived from surface water.

A relationship between increased water hardness and reduced visual quality aspects exists, as shown in table 2.5.

Table 2.5: Visual aspects of water quality and water hardness (n = 3073)

Water hardness category (meq/l)	% of total scores per water hardness category	
	Always clear and colourless category	Contains sometimes brown or black particles
0 - 2	88.7	3.1
2 - 4	85.2	7.9
4 - 6	84.3	7.9
$\geqslant 6$	76.8	8.7

The statistical significance of such a relationship between these two ordinal variables was verified by calculating the standard normal deviate (z), which follows a standard normal distribution if the null-hypothesis is true. According to a survey of calculated z-values given in appendix 2.1, the considered value of z equals 5.0 (one-tailed value of $p < 0.001$). Therefore the association between water hardness and water colour rating is indeed significant.

The same applies to the relationship between water hardness and the occurrence of brown or black particles. The association was calculated by means of the Kruskal-Wallis test (Siegel, 1956) which provides a measure of association (H) relating to the present combinations of ordinal and nominal variables. A survey of this type of associations is given in appendix 2.2, which shows that the association between water hardness and the occurrence of brown or black particles is significant ($H = 7.9$, $p = 0.005$). An increased tendency of the harder water to form deposits of for instance calcium and iron salts might be the cause of this effect.

A similar effect could be observed by comparing perceptible quality data with several physical-chemical data, relating to the drinking water supply of 50 communities, which were represented in the national inquiry by at least 20 observations. These data are described in appendix 2.3. Colour ratings varied from 1.00 - 1.61, the measured colour varied from 1 - 26 mg/l (Pt-Co) and the calcium concentration varied from 17 - 178 mg/l. With these 50 sets of data the following linear regressions were calculated:

Colour rating = 1.12 + 0.012 [colour] (mgPt/l) $r = 0.39$, ($\alpha < 0.01$)
and
Colour rating = 1.07 + 0.00205 [calcium] (mg/l) $r = 0.44$, ($\alpha < 0.01$)

These data indicate that the association between colour assessment by the consumer and water hardness is as important as the association with the analytically measured colour of the water at the water treatment station.

2.3.2 Temperature

About 80% of the individuals assessed water temperature as *"fresh"*. As can be expected water temperature is less frequently scored in the category *"fresh"* in case surface water is used as raw water source. The data for this temperature category, as presented in table 2.6 show a reduction from 82% for ground water supplies to about 71% for supplies with surface water as raw water source. As shown in appendix 2.2 the association between water temperature and water source is significant ($H = 50, p < 10^{-5}$)

Table 2.6: Water temperature assessment and type of raw water source (n = 3073)

Type of raw water source	% of total scores per raw water category		
	Fresh (cold)	Somewhat tepid	Tepid to warm
Ground water	82.3	17.5	0.2
Bankfiltrate	69.7	30.3	-
Stored surface water	71.9	27.7	0.3
Total (% of total)	78.6	21.2	0.2

Furthermore water temperature can be influenced during transport, particularly during the stay of water in the piping system in the houses. The data presented in table 2.7 confirm this hypothesis as within the category of one family houses or comparable houses which have shorter domestic plumbing systems the scores in the *"fresh"* category amount to 83% compared to 64% for the flat category.

Table 2.7: Water temperature assessment and type of house (n = 3073)

Type of house	% of total scores per house category		
	Fresh	Somewhat tepid	Tepid to warm
Bungalow or comparable	83.2	16.7	0.1
Flat	64.4	34.9	0.7
Others	83.2	16.8	0.0

According to the Kruskal-Wallis test (see appendix 2.2) temperature rating is associated in a significant way with the type of house ($H = 130, p < 10^{-5}$).

These data indicate that the effect on water temperature of the stay in the piping installation of flats is more pronounced than the effect of the type of raw water source.

2.3.3 Odour and taste

Taste of drinking water seems to be a sharper indicator of chemical contamination of water than smell. As table 2.8 shows water odour was scored in the categories *"good"* and *"foul"* by respectively 7.6% and 0.8% of the group sampled, while water taste was scored in these extreme categories by respectively 30% and 2.1% of the individuals. Generally about 7% of the persons rated the taste of water as offensive or worse, which value was only 2.6% for ground water supplies but amounted to 15% for surface water supplies. Comparable differences due to the water source were noticed for the flavour of tea prepared from tapwater, particularly for the case of water derived from bankfiltered surface water. The high degree of significance of these differences is further illustrated by the data of appendix 2.2 (H was respectively 210, 400 and 210 for the association between water source and respectively odour rating, taste rating and tea flavour rating).

Among the relatively small group of 3.2%, who found water odour offensive or worse, 88% did not object to the water smell when taking a bath or shower. Therefore the smell of water when used for a shower is much less critical than the smell of water when used for drinking purposes. This may be partially due to the masking effect of the use of perfumed soap.

Table 2.8: Odour (O) and taste (T) assessment of drinking water (n = 3073) and of tea* (n = 2766) in relation to the type of raw water source

Type of raw water	Odour and taste assessment (% of total per raw water category)												
	Drinking water										Tea flavour		
	Good		Not perceptible		Faint not offensive		Offensive		Foul		Good	Could be better	Not good
	O	T	O	T	O	T	O	T	O	T			
Ground water	10.5	38.9	85.7	52.7	3.0	5.8	0.6	1.8	0.2	0.8	89.8	8.8	1.4
Bankfiltrate	3.4	13.1	74.3	45.1	12.0	24.0	8.6	11.4	1.7	6.3	54.4	37.0	8.5
Stored surface water	2.0	13.4	82.1	54.0	8.6	18.0	5.4	10.2	1.9	4.4	72.6	21.3	6.1
Total (% of total)	7.6	30.1	84.0	52.6	5.2	10.4	2.4	4.8	0.8	2.1	82.7	14.1	3.2

*Included only those individuals who actually consume tea

In order to examine possible effects of water hardness on water taste assessment both parameters were compared, based on the data for ground water supplies. Only ground water supplies were considered in order to eliminate interfering effects due to the type of the water source.

Table 2.9 shows the negative effect of increasing water hardness on the taste of water as well as on the taste of tea. In general the data of appendix 2.1, although including surface water supplies, confirm the described association.

The taste assessment of tea prepared from hard tapwaters might be partially

Table 2.9: Hardness of ground water supplies and taste assessment of water (n = 1966) and of tea (n = 1767)

Hardness category (meq/l)	% of total per hardness category							
	Water taste					Tea taste		
	Good	Not perceptible	Faint not offensive	Offensive	Foul	Good	Could be better	Not good
0 - 2	52.1	44.6	2.8	0.5	0.0	97.6	2.1	0.3
2 - 4	32.7	60.4	6.3	0.5	0.1	91.0	7.7	1.3
4 - 6	38.5	48.6	6.8	4.1	2.0	84.1	13.3	2.6
$\geqslant 6$	31.2	57.6	8.0	2.2	1.0	83.1	15.3	1.6

influenced by the formation of a skin on the tea surface, which effect is more evident when using hard water, as indicated by table 2.10. Comparison with the data of appendix 2.2 shows that water hardness is associated with skin formation on the tea in a significant way (H = 270, $p < 10^{-5}$), but no significant association between water hardness and the offensiveness of this skin could be shown. 6% observe an offensive skin in the case of soft waters (0 - 2 meq/l hardness), which value increases to 37.8% for the category of hard waters (6 meq/l or more)

Table 2.10: Water hardness and skin formation on tea (n = 2766)

Hardness category (meq/l)	% of total per hardness category	
	Observing a skin on the tea	Indicating the observed skin on tea as offensive
0 - 2	11.5	6.1
2 - 4	34.3	20.2
4 - 6	53.9	32.5
$\geqslant 6$	58.6	37.8
Total (% of total)	42.2	25.7

A further evaluation of the contribution of water hardness to water taste has been made by carrying out correlation calculations for 50 communities, as presented in appendix 2.3. The following results for the relation between the parameters calcium and magnesium and the calculated average taste rating per community were obtained:

$$\text{Taste rating} = 1.53 + 0.0066 \; [Ca] \; (mg/l) \qquad r = 0.48 \; (\alpha < 0.001)$$

and

$$\text{Taste rating} = 1.50 + 0.062 \; [Mg] \; (mg/l) \qquad r = 0.61 \; (\alpha < 0.001)$$

while

$$[Mg] \; (mg/l) = 0.31 + 0.12 \; [Ca] \; (mg/l) \qquad r = 0.86 \; (\alpha < 0.001)$$

These data suggest a particular role of magnesium in taste impairment of drinking water in The Netherlands.

Another factor influencing water taste assessment could be the age of the water distribution system as shown in table 2.11. Such an association between age of the house and water taste was not confirmed statistically (see appendix 2.2). On the other hand a significant association (H = 16, p = 0.004) was found between the taste rating and the type of material used for domestic plumbing.

Table 2.11: Year of construction of the house and water taste assessment (n = 3073)

Taste category	% of total per taste category	
	Year of construction	
	Before 1950	1950 or later
Good	42.0	58.0
Not perceptible	37.3	62.7
Faint not offensive	43.4	56.6
Offensive	49.3	50.7
Foul	53.8	46.2

Furthermore appendix 2.1 shows the highly significant association between taste rating and respectively odour rating (z = 23, p<0.001), tea flavour rating (z = 21, p<0.001) and water temperature (z = 11, p<0.001)

The results of the assessment of the quality of the perceived taste and odour of drinking water in The Netherlands are given in table 2.12. As relatively few observations are available and subjects had to use prescribed qualifications for the taste and odour, the data obtained should only be considered as indicative.

Table 2.12: Raw water source and qualification of the perceived water odour (O) (n = 98) and water taste (T) (n = 211)

Raw water source	% of total per water source category									
	Qualification									
	Chlorine-like		Earthy		Putrid		Metallic		Faint	
	O	T	O	T	O	T	O	T	O	T
Ground water	20	32	20	22	7	2	20	8	33	36
Bankfiltrate	44	36	33	29	0	3	6	16	17	16
Stored surface water	60	43	20	22	5	2	5	9	10	24
Total (% of total)	51	39	23	23	4	2	7	10	15	26

The odour and taste of drinking water derived from ground water which had an offensive or foul taste or smell mostly obtained the qualification *"faint"*, while in the case of the waters derived from surface water the qualification *"chlorine-like"* was most frequently used.

2.4 Aspects possibly affecting sensory water quality assessment

2.4.1 Knowledge of raw water source and opinion on water taste elsewhere

As water quality assessment might be affected by the knowledge of the source used for the local water supply as well as the opinion on the taste of local drinking water in comparison to the taste of drinking water elsewhere, such questions were included in the questionnaire. The results are given in table 2.13.

Table 2.13: Opinion on raw water source and on water taste elsewhere in relation to the actual raw water source (n = 3073)

Raw water source	% of total per raw water category							
	Opinion on							
	Raw water source			Water taste elsewhere is				
	Ground water	Surface water	Unknown	Better	Same	Sometimes better sometimes worse	Worse	No opinion
Ground water	55.1	17.1	27.8	4.9	25.1	32.7	15.8	21.6
Bankfiltrate	25.1	40.6	34.3	23.6	22.1	13.3	19.6	21.4
Stored surface water	15.7	56.2	28.1	21.6	12.8	19.4	29.4	16.7
Total (% of total)	42.0	29.8	28.2	11.3	23.6	17.7	26.2	21.2

Table 2.13 shows that about 80% of the consumers express an opinion on the relative taste quality of their drinking water. In the areas supplied from ground water 4.9% scored water taste elsewhere as better, which figure amounts to 23.6% for areas supplied from bankfiltered water.

The data of table 2.13 further indicate that about 55% of the people consuming water derived from ground water or from stored surface water are aware of the actual water source. As bankfiltrate is a mixture of surface water and ground water it is not surprising that consumers within this category are somewhat confused about this item, although 40% still indicated surface water as the source of water.

Approximately 50% of the consumers are well informed about the type of raw water used for the water supply. Therefore the possible effect of this factor on water quality assessment needs further consideration. In table 2.14 taste assessment by consumers of drinking water actually derived from ground water or from stored surface water is compared with the taste assessment of consumers expecting the raw water source to belong to one of these two categories.

Table 2.14: Taste assessment of water derived from ground water (n = 1999) or stored surface water (n = 894) and of water expected to be derived from these two sources (n = 1291, n = 915)

Raw water category	% of total per raw water category				
	Taste category				
	Good	Not perceptible	Faint not offensive	Offensive	Foul
Ground water					
- actual	38.8	52.8	5.8	1.8	0.8
- expected	39.7	50.7	7.0	1.3	1.3
Stored surface water					
- actual	13.8	52.9	18.6	10.3	4.4
- expected	18.6	53.0	17.0	8.4	3.0

The data given in table 2.14 do not suggest any consistent effect of the expected type of water source on the taste ratings.

2.4.2 Odour of the air at the residence and at the place of work

It was supposed that individuals perceiving odorous air contamination at their residence or at the place of work could be less sensitive for water taste and odour, for instance as a result of adaptation or habituation phenomena. The results of questions on these factors are shown in table 2.15. The number of observations for odour quality at the place of work concerned only 1514 values as about 50% of the consumers did not work outside their residence.

The data do not confirm the hypothesis, as consumers exposed to odorous air pollutants generally also assessed the water taste more often as offensive.

2.5 Perceptible water quality aspects and general water quality assessment

2.5.1 Acceptability rating of water quality

It was anticipated that acceptability of water quality would be mainly dependent on the judgements relating to the perceptible water quality aspects. As shown in table 2.16 about 3% of the individuals rated the water quality as *"sometimes disliked"* or worse.

The acceptability rating is associated in a significant way with perceptible water quality aspects, as is shown by the data of appendix 2.1. The z-values for the association with acceptability rating were respectively 35 for taste rating, 25 for tea flavour rating, 20 for odour rating, 16 for colour rating and 11 for temperature rating (see also appendix 2.3).

From these data it may be concluded that the acceptability rating of water quality is mainly based on the taste assessment.

Table 2.15: Odour assessment of air at the residence (n = 3073) and at the place of work (n = 1514) in relation to water taste rating

Odour assessment	% of total per air quality category					% of total
	Water taste assessment					
	Good	Not perceptible	Faint not offensive	Offensive	Foul	
At the residence						
Very satisfactory	39.6	50.5	6.5	2.1	1.4	37.7
Satisfactory	25.3	57.9	11.4	4.1	1.4	31.2
Usually not unsatisfactory	26.0	51.2	13.7	6.3	2.9	23.8
Unsatisfactory, sometimes offensive	16.4	47.9	15.5	13.3	7.2	5.4
Very unsatisfactory, often offensive	13.6	37.3	18.6	25.4	5.1	1.9
At the place of work						
Pleasant	36.9	51.7	6.6	4.1	0.7	27.2
Usually good	24.8	58.7	11.0	4.8	0.7	39.7
Sometimes offensive	26.1	53.0	13.5	5.1	2.4	22.1
Regularly stinking	26.2	51.5	14.6	5.8	1.9	6.8
Always stinking	23.4	40.6	18.8	10.9	6.3	4.2

Suggested causes of bad odour of ambient air (n = 955)

Industrial	54.4
Agricultural	12.9
Traffic	20.3
Ditch water etc.	9.0
Other sources	12.4

Table 2.16: Acceptability assessment of water quality and judgements on water taste (n = 3073)

Category of acceptability of drinking water quality	% of total per column					% of total
	Taste category					
	Good	Not perceptible	Faint	Offensive	Foul	
Excellent	58.1	18.4	3.8	0.7	0.0	7.6
Satisfactory	38.2	66.3	35.9	8.9	1.5	84.1
Not complaining	3.5	12.8	47.2	30.1	10.8	5.1
Sometimes disliked	0.2	0.7	8.4	36.3	26.2	2.4
Drinking only if no choice	0.0	1.8	4.7	24.0	61.5	0.8

2.5.2 Safety rating of water quality

Among the tested group about 1% indicated water quality to be unsafe from a public health point of view. Like in the case of acceptability, water taste is an important factor in the subjective safety assessment (table 2.17).

Table 2.17: Assessment of safety of water quality and judgements on water taste (n = 3073)

Category of safety of drinking water	% of total per column					% of total
	Taste category					
	Good	Not perceptible	Faint	Offensive	Foul	
Safe	97.6	90.0	75.6	51.3	44.3	88.0
Doubt safety sometimes	2.1	9.2	22.5	44.5	40.1	10.8
Not safe	0.3	0.8	1.9	4.2	15.7	1.2

From the data of appendix 2.1 it may be concluded that the safety rating is more strongly associated with the acceptability rating ($z = 20$) than with any of the perceptible aspects studied.

2.6 Conclusions

An inquiry held in June 1976 among 3073 individuals of 18 years and over, living in 89 communities in The Netherlands, showed that:

23 % observed an offensive skin on the tea;
6.9% rated the taste of their drinking water as offensive or worse;
3.2% disliked the taste and odour of the home brewed tea;
3.2% rated drinking water odour as offensive or worse;
1.3% rated drinking water colour and turbidity as offensive or worse;
0.4% indicated that water odour was offensive taking a bath or a shower;
0.2% rated the temperature of their tapwater as tepid or worse.

Generally drinking water derived from ground water was significantly better in relation to all aspects of sensory assessment of water quality compared to water prepared from surface water (appendix 2.2).

It was found that increasing hardness generally corresponds to reduced visual quality of water, such as more colour and turbidity, higher incidence of brown of black particles (table 2.5), higher incidence of an offensive skin on the tea (table 2.10) and reduced taste of water and tea (table 2.9). Particularly increased levels of magnesium seem to affect the taste of water derived from ground water.

The detention of drinking water in the distribution system could be shown as a significant factor in reduced freshness of tapwater. Only 64% of the persons living in flats rated the water as fresh against 83% of the individuals living in one family houses or comparable houses (table 2.7). The effect of the stay of the water in the distribution network showed a higher association with reduced coldness of the water than the effect of the type of water source (H-values were respectively 130 and 50).

Water taste was rated in the categories *"good"* and *"not perceptible"* by 92% of the consumers of water derived from ground water as compared to a value of only 58% for consumers supplied by means of bankfiltered water (table 2.8). Water taste rating was associated in a significant way with water odour rating ($z = 23$), tea flavour rating ($z = 21$) and water temperature ($z = 11$) (appendix 2.1). The most frequently used description of the quality of the taste and odour of water derived from surface water sources was *"chlorine-like"* (table 2.12).

No consistent effects of the expected type of water source on the test results could be demonstrated (table 2.14). Furthermore the data were not suitable for the evaluation of a possible effect of air pollution by offensive odours on the sensory assessment of water quality (table 2.15).

Water taste could be shown to be the major factor determining the acceptability assessment of water quality. Acceptability rating was more strongly associated with the opinion on the safety from a public health point of view than any of the aspects perceptible by the senses. 3.2% of the tested group rated acceptability as *"sometimes disliked"* or worse (table 2.16) and 1.2% qualified their drinking water as *"unsafe"* (table 2.17).

In order to identify more specific causes of reduced water quality, as perceived by the chemical senses, individual types of drinking water have to be studied. The results of such investigations are discussed in the following chapters.

Water-sniffing

34

3 Sensory assessment of 20 types of drinking water by a selected panel

3.1 Introduction

Judgements of local tapwater quality by subjects, who are used to its taste, might be rather different from judgements by non-residents used to other tapwater qualities. Therefore a studie was carried out in which 20 types of drinking water were presented to a group of persons living in different parts of The Netherlands. In this way a more objective and specific comparison of the sensory aspects of water quality is possible than in the case where a comparison would only be based on the general data of the inquiry as described in Chapter 2. Such a comparison will provide the necessary basis for the subsequent identification of chemical substances in drinking water which are responsible for the adverse taste and smell of particular types of tapwater.

In this chapter the results of the sensory assessment of the quality of 20 types of drinking water are described whereas in subsequent chapters the types of chemical compounds present and their relation to water taste are discussed.

This chapter first describes the selection procedure and several characteristics of the selected panel, the stimulus presentation and test conditions, the type of psychological scale and the selection of the 20 types of drinking water. Finally the results obtained are discussed.

3.2 The panel

3.2.1 Panel selection

A panel had to be selected which was as representative as possible in relation to the taste attitude of the population of The Netherlands. Because of practical reasons the total number of panel members however, had to be limited to about 50 and a minimum age had to be fixed, for which 18 years was chosen. Anosmic or very insentive subjects had to be excluded.

An invitation to participate as a panel member, during three sessions on Saturdays, was sent to members of 1900 households in The Netherlands by a contracted institute (NIPO, Nederlands Instituut voor de Publieke Opinie en het Marktonderzoek b.v., Amsterdam). This sample of the population, taken aselectively as far as province and community size are concerned, was composed according to the distribution of the housing stock and stratified according to the number of houses per community. However, the relatively small number of approximately 200 positive responses cannot be considered as a representative sample of the national population. From this group 148 individuals were selected on basis of the domicile distribution.

This group was submitted to a test concerning the odour sensitivity for aqueous solutions of isoborneol and o-dichlorobenzene. Isoborneol has an earthy, musty smell, which represents certain water odours of biological origing and o-dichlorobenzene has a chemical, disinfectant-like smell, which was considered to be a representative industrial contaminant of water. The subjects were tested for odour sensitivity as it was expected that the presence of odorous organic micropollutants in drinking water would be mainly responsible for impaired taste.

The sensitivity test was carried out at the homes of the individuals by offering them concentrations of the odorants in a fixed order as indicated in table 3.1. A fixed order was prescribed to ensure as much as possible comparable results, although the tests were carried out by 10 different instructors and under different local circumstances.

Table 3.1: Order of presentation of duplo concentrations of aqueous solutions of isoborneol and o-dichlorobenzene.

Isoborneol (µg/l)				o-Dichlorobenzene (µg/l)			
No.	Conc.	No.	Conc.	No.	Conc.	No.	Conc.
1	0.3	7	1.5	13	450	19	3
2	45	8	150	14	150	20	150
3	4.5	9	4.5	15	3	21	45
4	1.5	10	0.3	16	15	22	15
5	15	11	15	17	45	23	1500
6	150	12	45	18	1500	24	450

The forced choice method was used. During each presentation the flask containing the odorant was accompanied by three flasks containing blanks. The subjects had to indicate orally to an instructor, which of the four flasks contained the odorant. The place of the flasks was varied at random, whilst the four flasks were coded with the letters M, D, V and T. All solutions were prepared locally, using The Hague tapwater and Erlenmeyer flasks, covered with a watch-glass. The flasks had to be shaked manually before smelling. The criterion for setting the lowest perceptible concentration, the individual threshold, was a consistent correct response at decreasing concentrations in both series of the duplo experiment. Those cases, where a correct and a wrong indication for a certain concentration in the duplo experiment were directly followed at the next lower concentration by two correct scores, were also accepted as correct scores.

Preferably those subjects were included in the panel which showed a higher sensitivity than the average sensitivity for both compounds while those persons were excluded who could not detect the highest offered concentration of one of the two odorants. After selection for sensitivity, age, sex and domicile a group of 56 subjects resulted for the taste experiments. Table 3.2 gives the distribution of the domiciles over the Dutch provinces for the members of the panel, as well as for the 148 subjects tested and the total population of The Netherlands.

Figure 3.1 gives the distribution of individuals within the panel, the subjects tested and the national population (CBS, 1976) as far as age and sex are concerned.

The odour sensitivity distribution of the panel, compared to the group of 148 individuals tested is presented in figure 3.2 for isoborneol as well as o-dichlorobenzene.

Table 3.2: Domicile distribution for the Dutch population (CBS, 1976), the persons tested and the panel.

Province	% of the population (n = 13.700.000)	% of the persons tested (n = 148)	% of the panel (n = 56)
Groningen	3.9	2.8	3.6
Friesland	4.1	6.2	5.4
Drente	3.0	2.1	3.6
Overijssel	7.2	6.9	7.1
Gelderland	11.9	10.3	10.6
Utrecht	6.3	6.9	7.2
Noord-Holland	16.7	13.2	16.0
Zuid-Holland	22.2	26.9	23.1
Zeeland	2.4	4.1	3.6
Noord-Brabant	14.3	15.1	14.2
Limburg	7.7	5.5	5.4

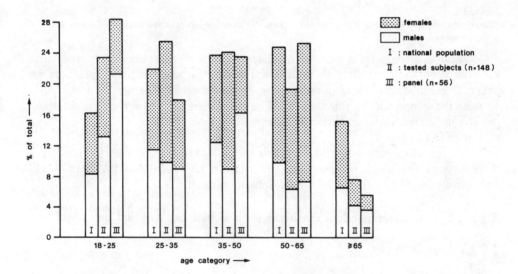

Figure 3.1
Distribution of age and sex of the population in the Netherlands, the tested subjects and the panel.

These data indicate that the domicile distribution of the panel was rather representative for the Dutch population but that relatively many males in the age category 18-25 years and relatively many females in the age category 45-65 years were selected. Furthermore the age category 18-25 years was over-represented and relatively few subjects were represented in the age category of 65 years and over, due to the limited number of subjects available for the selection of the panel.

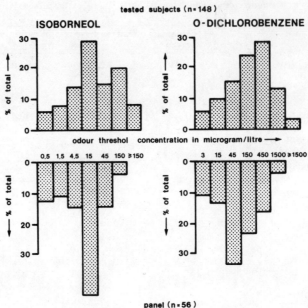

Figure 3.2
Odour sensitivity distribution of the tested subjects and the panel for aqueous solutions of isoborneol and o-dichlorobenzene.

Additional data on a number of characteristics of the panel compared to the national population are given in table 3.3. For this purpose the panel members were asked to complete the same questionnaire as described in Chapter 2 in the period between the second and the third session. The panel members showed a somewhat more positive judgement on water taste and acceptability than the sample of the population, described in Chapter 2.

The panel members obtained a bonus of Dfl 200 in case they participated in all three sessions. During the three sessions held on March 27, May 22 and August 28, 1976 respectively 54, 51 and 52 subjects were actually present.

3.2.2 Factors affecting odour sensitivity for o-dichlorobenzene and isoborneol

3.2.2.1 *General aspects*

Besides the determination of the OTC for aqueous solutions of o-dichlorobenzene and isoborneol during the selection procedure, the OTC was also measured during each of the three successive panel sessions. During the panel sessions the odour sensitivity was tested in the same way as described in paragraph 3.2.1, only the Erlenmeyer flasks were replaced by beakers, covered with a watch-glass. The test was carried out just before and just after lunch. The odour sensitivity data were used for detection of variations in the average sensitivity of the panel members during the three sessions.

Based on these data also other aspects can be studied like the influence of sex, age and smoking on the odour sensitivity for the two compounds considered.

Table 3.3: Comparison of some characteristic figures for the participating panel members and the sample of the national population (18 years and over) involved in the national inquiry (see Chapter 2).

Aspect	Category	Panel (n = 52)	Sample of the national population (n = 3073)
Average quantity of water directly consumed as drinking water (ml/person/day)		245	233
Assessment of local water taste (% of scores)	Good	33	30
	Not perceptible	48	53
	Faint	15	10
	Offensive	4	5
	Foul	0	2
Acceptability of local water quality (% of scores)	Excellent	36	28
	Satisfactory	46	51
	Not complaining	14	14
	Sometimes disliked	4	4
	Drinking if no choice	0	4
Type of water sources	Surface water*	40	35
	Ground water	60	65

*including mixtures of surface water and ground water.

In the following paragraphs the effect of these factors on odour sensitivity will be discussed in order to find out which factors are most relevant as selection criteria for panels to be used in sensory assessment of drinking water quality.

3.2.2.2 *Variability of individual odour threshold in time*

The average odour sensitivity of the panel on the three session days was respectively 10, 6 and 8 µg/l for isoborneol solutions and 300, 170 and 200 µg/l for o-dichlorobenzene solutions. These data suggest that the average odour sensitivity of the panel was higher during the second and third session than during the first session. As the respective t-values were for isoborneol -1.90 and for o-dichlorobenzene -3.35 this tendency was nearly significant for isoborneol and significant for o-dichlorobenzene. ($|t| = 1.96$ for $\alpha = 0.05$).

In accordance with this finding table 3.4 indicates that more subjects showed a stable odour sensitivity during the three sessions for isoborneol than for o-dichlorobenzene.

Table 3.4: Intra-individual changes in odour sensitivity of panel members during three sessions.

Aspect	% of panel members	
	isoborneol	o-dichlorobenzene
Same threshold in all three sessions	24	2
Same threshold in two sessions	52	66
Different threshold in all three sessions	24	32

For the majority of subjects the odour threshold for both compounds changed not more than a factor 3 during the three sessions. Changes of a factor 30 or more in the OTC value between the first and the third session occurred for 10% of the subjects in relation to o-dichlorobenzene and were not found for isoborneol. The intra-individual variance in these OTC values seems to be larger than the variance in taste thresholds for sodium chloride solutions (Mefferd, Wieland, 1968).

In paragraph 3.5.4 the possible effect of changes in the odour sensitivity on *"taste"* assessment of drinking water will be considered.

3.2.2.3 *Sex and odour sensitivity*

Based on the data obtained during the selection procedure and on the data relating to the panel sessions, a possible relation between sex and odour sensitivity was investigated. As the panel data were based on more observations per subject and were obtained under more carefully controled conditions, the threshold values for the panel members are considered of primary importance. The data relating to the larger group of 148 subjects are used for confirmation purposes.

As shown in appendix 3.1 a number of multiple linear regressions have been calculated to investigate the statistical significance of possible associations between individual characteristics and odour sensitivity. The data applying to the panel indicate a possible relationship between sex and odour sensitivity for isoborneol ($t = 1.77$). The possible higher sensitivity of females for isoborneol was also found for the larger group from which the panel has been selected ($t = 1.81$), although the available data are insufficient to prove the statistical significance of such a relationship. For both groups no indication of an effect of sex on odour sensitivity for o-dichlorobenzene was found.

These results indicate, that the greater acuity of females for bitter tasting compounds like quinine (Soltan and Bracken, 1958), (Kaplan and Fischer, 1965) is not found for the two odorants studied. As sex differences in odour sensitivity for compounds such as pyridine and m-xylene were found to be rather small (a factor 1.5-2) (Koelega and Köster, 1974), larger groups or more accurate determinations are needed to prove such possible sex differences in odour sensitivity for the compounds studied.

3.2.2.4 *Smoking and odour sensitivity*

Smoking is reported to reduce taste sensitivity (Arfmann and Chapanis, 1962) (Kaplan *et al.*, 1964) (Fischer, 1971). Therefore the effect of smoking on odour sensitivity has been studied for the two odorants.

The intensity of cigarette smoking of the subjects tested is shown in table 3.5. No other smokers than cigarette smokers were present in this study. Table 3.5 indicates that panel members, selected partly on basis of their odour sensitivity, were smoking less cigarettes than the larger group of 148 subjects. Because more than half of both groups consisted of non-smokers, all subjects who smoked one cigarette or more per day were considered to be smokers.

The t-values, listed in appendix 3.1, indicate a possibly reduced odour sensitivity for o-dichlorobenzene due to smoking (for the panel $t = 1.23$, for the selection group $t = 1.85$), although these values are statistically not significant. In this case the two-tailed value of t at 5% probability of exceedence is ±1.96. For isoborneol no effect of smoking on odour sensitivity could be shown.

Table 3.5: Distribution of intensity of cigarette smoking in the tested group of subjects and in the panel.

Number of cigarettes/day	% of panel members (n=56)	% of group tested (n=148)
0	58	51
1-9	14	11
10-19	19	25
20 and more	9	13

3.2.2.5 *Age and odour sensitivity*

According to Hughes (1969) the senses of taste and smell show a progressive reduction in sensitivity with increasing age, particularly from the age of 60 years onwards. Kaplan *et al.*, (1964, 1965) found no age-related differences in taste sensitivity for quinine and 6-n-propylthiouracil for non-smokers, 16 to 55 years of age. The data of appendix 3.1 do not show an effect of age on odour sensitivity for isoborneol or o-dichlorobenzene.

3.2.2.6 *Water source and odour sensitivity*

A rather unexpected relationship was found between odour sensitivity for isoborneol and possibly too for o-dichlorobenzene and the type of water source to which the subject is normally exposed. The t-values for the panel data were respectively 2.66 and 1.65. Only for isoborneol this effect of the water source on odour sensitivity was confirmed by the data of the larger group used for selecting the panel.

A possible explanation for this effect could be that subjects which are used to the presence in their drinking water of compounds such as isoborneol and dichlorobenzene are inclined to consider low odour intensities of these compounds as *"background noise"* (see also paragraph 1.2.2.2). Another possible explanation consists of a reduced sensitivity due to the more frequent exposure to odorous ambient air of subjects living in the larger cities. However, such a relationship was not confirmed by the data of appendix 3.1. No significant associations were found for odour sensitivity and community size.

3.2.2.7 *Conclusions and recommendations*

The results obtained indicate that for isoborneol and o-dichlorobenzene intra-individual changes in odour sensitivity over longer periods of several weeks are generally not greater than a factor 3, while this variability differs for different compounds. Odour sensitivity for isoborneol is less variable than for o-dichlorobenzene. Changes in the average sensitivity of a panel do occur and should be recorded, using odorants which are as representative as possible for the taste and odour of the water samples studied. In this way the comparability of results obtained during different panel sessions can be improved.

The sex of panel members was not found to influence significantly the odour sensitivity for o-dichlorobenzene and isoborneol. Also other studies (Koelega and Köster, 1974) showed only small differences in odour sensitivity between males and females for compounds which can be present in drinking water. Although indications exist that females will be somewhat more sensitive panel members than males, it seems to be more important to control for intra-individual variability than for sex, during the selection of a panel for water quality assessment.

Smoking could not be shown to reduce the odour sensitivity of subjects, although the data indicate a possible effect of smoking on odour sensitivity for o-dichlorobenzene. The possible minor effects of smoking on odour sensitivity should only be taken into account during panel selection in case large groups of subjects are available.

As the water source was significantly associated with the odour sensitivity for isoborneol, subjects supplied with drinking water derived from ground water should be prefered in panels which have to detect odorous compounds in surface water supplies.

Finally it can be stated that a preselection of members of a panel for water quality assessment is always needed as far as odour sensitivity for a compound such as o-dichlorobenzene is concerned. In this way anosmic and very insensitive subjects can be excluded. Secondly attention should be paid to intra-individual variability in odour sensitivity and to the water source of the subjects. Factors such as sex, smoking and age are generally only of minor importance.

3.3 Stimulus presentation, test conditions and psychophysical scaling

The panel sessions were held in two adjacent rooms in one of the buildings of the University of Utrecht. The panel members arrived at 10 a.m. and were first offered coffee. At 10.15 hours they were seated behind each other along the walls of each room, to minimize mutual influencing. During the experiment no conversation was allowed.

After a short introduction the stimuli were presented in transparent and odourless polythene cups (Thovadec) containing 15-25 ml of drinking water and coded with successive figures. The cups were filled just before presentation from larger sampling bottles, which had stayed overnight at room temperature.

The session started with the presentation of two cups of local tapwater for rinsing purposes, followed by a series of 40 stimuli, composed of 8 types of drinking water in randomized order. At a sound signal the sample had to be smelt and tasted, without swallowing it. To minimize adaptation effects, the time interval between two presentations was set at 60 seconds. After a coffee break, a lunch interval and a tea break the same series of 2 blanks and 40 stimuli were presented. During the third session the afternoon was used for other experiments, which means that during the third session the 8 tapwaters were presented to the panel only 10 times instead of 20 times. At 16.00 hours the experiments were finished.

The panel members were asked to indicate on a preprinted punch card which item out of five reflected most closely the perceived smell and taste of the water. The used taste items are indicated in table 3.6. The odour items were the same except that the word "taste" or "tastes" was replaced by "smell" or "smells".

Since this type of scale only allows the use of an ordinal classification, an interval scale was constructed by the use of the method of triads (Torgerson, 1967). All possible combinations of 3 out of the 5 items were presented to the panel members, which had to indicate which pair was most similar and which pair was most dissimilar. Every combination of 3 items was presented 6 times in randomized positions during the second panel session. The obtained data were treated by multidimensional analysis using the *"minissa-1"* computer program (Roskam, 1970). A 1-dimensional as well as a 2-dimensional solution was obtained.

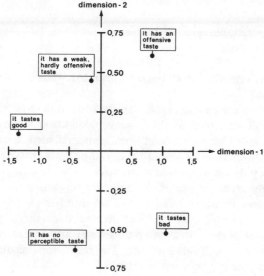

Figure 3.3
2-dimensional scale values for the descriptive terms of the taste [odour] scale.

The results of the 2-dimensional solution (see figure 3.3) showed that the items 3 and 4 scored high positive and the items 2 and 5 high negative. Apparently the subjects made a distinction between the good-neutral-bad-dimension and the offensive-dimension.

For computing purposes the 1-dimensional solution was used which resulted in the scale values as indicated in table 3.6. Both solutions had a satisfactory fit although the 2-dimensional solution was slightly better.

Table 3.6: Items used for taste description and corresponding scale values.

Item number	Original description	Translated english description	Scale value*
1	Het smaakt lekker	It tastes good	0.00
2	Het heeft geen waarneembare smaak	It has no perceptible taste	0.74
3	Het heeft een zwakke, weinig hinderlijke smaak	It has a weak, hardly offensive taste	1.41
4	Het heeft een hinderlijke smaak	It has an offensive taste	2.07
5	Het smaakt vies	It tastes bad	2.87

*Note that a higher scale value corresponds with a lower quality.

Panel members were also asked to describe on a separate sheet the quality of the perceived taste for two series of 8 tapwater samples during a session. One of the following items had to be indicated:

- It has a chlorine-like taste
- It has an earthy taste
- It has a putrid taste
- It has a metallic taste
- It has a faint, unqualifiable taste
- It has no perceptible taste

3.4 Selection and sampling of 20 types of drinking water

As this study is mainly aiming at the identification of the role of organic substances in impaired taste and odour of drinking water, particularly those supplies in The Netherlands were included which are using contaminated surface water as a raw water source. Furthermore a number of ground water supplies were selected as references, while a particular supply using ground water and a supply using surface water were represented in all three series, to provide a basis for comparison as well as an indication of seasonal fluctuations in water quality. A total number of 20 cities with different supplies was selected in this way and was divided over three series. Each series contained drinking waters derived from ground water and surface water and drinking waters with low as well as high salt contents. The framework resulting from the above mentioned considerations is presented in table 3.7.

To enable the study of associations between water quality data and taste assessment data, water sampling had to be carried out in such a way that a composite sample for a certain community or group of communities was collected in case more than one pumping station was supplying the area. For this purpose stainless steel vessels were designed which could be filled stepwise with water at different sampling locations without allowing air contaminants to come into contact with the water. The stainless steel vessels, which had a content of 200 litres, were filled with nitrogen gas, after careful cleaning with acetone, hydrochloric acid, sodium hydroxide solution and tapwater rinsing steps. During sampling the nitrogen gas was allowed to leave the vessel via a column packed with activated carbon powder. The tubing by which the vessel was connected to the tap also consisted of stainless steel.

Sampling locations were selected after consultation with waterworks personnel. People responsible for sampling were instructed to flush the tap prior to sampling. The vessels were transported by car to the sampling locations and finally to the laboratory where they were emptied by applying pressure with nitrogen gas.

The water from these vessels was used for different types of chemical determinations as will be described later. Samples for analysis of volatile organic compounds were collected separately in 1 litre glass bottles.

The composite water samples from the above mentioned sampling points were also collected in carefully cleaned glass bottles of 25-50 litres, which were delivered at the University of Utrecht the evening before the panel session. Glass bottles were chosen to prevent any possibility of contamination, as the panel experiment could not be repeated

Table 3.7: Survey of selected types of drinking water and their general characteristics.

Name of community	Number of pumping stations	Main type of raw water source	Type of storage facilities	Oxidative treatment with	Conductivity class 4)
1st series, sampling date March 26, 1976					
Apeldoorn	4	Ground water	-	-	1
Barendrecht	1	Ground water	-	-	3
Dordrecht	2	Rhine river	Reservoir	O_3/Cl_2	3
's-Gravenhage	1	Rhine river	Dune infiltr.	-	3
Hardinxveld	1	Rhine river	Bank filtr.	-	3
Leiden	1	Canal water	Dune infiltr.	O_3/Cl_2	3
Rotterdam	2	Meuse water	Reservoir	Cl_2	2
Utrecht	4	Ground water	-	-	1
2nd series, sampling date May 21, 1976					
Alkmaar	2	Rhine river	Dune infiltr.	Cl_2	3
Amsterdam	2	Rhine river	Dune infiltr.	Cl_2	2
Dordrecht	2	Rhine river	Reservoir	O_3/Cl_2	3
Enkhuizen/Hoorn	1	IJssel lake	Reservoir	Cl_2	3
Groningen	2	Ground water 1)	-	-	1
Leeuwarden	1	Ground water	-	-	2
Schoonhoven	1	Rhine river	Bank filtr.	-	3
Utrecht	4	Ground water	-	-	1
3rd series, sampling date August 27, 1976					
Arnhem	1	Ground water	-	-	1
Dordrecht	2	Rhine river	Reservoir	O_3/Cl_2	3
Eindhoven	1	Ground water	-	-	1
Enschede	2	Twenthe canal	Dune infiltr. 2)	-	2
's-Hertogenbosch	1	Ground water	-	-	2
Utrecht	4	Ground water	-	-	1
Zwolle	1	Rhine river	Bank filtr.	O_3/Cl_2 3)	2
Zwijndrecht	1	Rhine river	Bank filtr.	O_3	3

1) A minor quantity is derived from surface water
2) System resembles most closely that of dune infiltration
3) 25% of the flow is treated with ozone
4) Conductivity classes in $\mu S/cm$ (20°C) : 1: $\leqslant$ 400, 2:401-750, 3: >750

in case of a failure due to contamination of the sampling device.

All sampling of the 8 types of drinking water to be studied during each session day was carried out the day before the panel session. The samples to be analysed for volatile compounds, were stored 1-5 days at 4°C. All other water samples were processed for the determination of organic compounds, directly after arrival at the laboratory.

3.5 Taste and odour assessment of 20 types of drinking water

3.5.1 Relation between taste and odour rating of drinking water

Based on the taste and odour ratings obtained during the first session, as shown in figure 3.4, it can be concluded that smell of drinking water is a much less pronounced quality aspect than water taste. For the taste and odour ratings, as presented in figure 3.4, the Pearson correlation coefficient amounts to 0.42 ($\alpha = 0.001$).

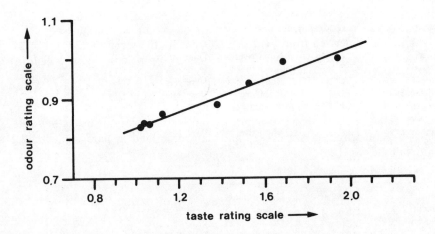

Figure 3.4
Relation between taste rating and odour rating for the 8 types of drinking water studied during the first session of the panel.

The taste ratings for the 8 types of drinking water considered, varied between values from 1.0-2.0 on the scale as given in table 3.6, while the odour ratings for the same samples varied only between 0.85-1.00. For the water with the highest rating on the scales 33% of the subjects scored in the *"it tastes bad"* category while only 1% scored in the *"it smells bad"* category. This illustrates the rather poor suitability of the determination of the smell of water as compared to the determination of its taste.

Therefore only the results of the taste assessment of water samples have been discussed in the following paragraphs.

3.5.2 Results of the taste rating

After comparison of the data obtained during the panel sessions it was found that in relation to the two types of tapwater which were present at all three session days, large differences were detected during the experimental period as shown in table 3.8.

Table 3.8: Observed mean taste ratings for the two tapwaters studied in all three sessions

Type of source of tapwater	Mean taste scale value		
	1st session	2nd session	3rd session
Ground water	1.01	1.20	1.18
Surface water	1.68	1.95	2.40

The drinking water derived from surface water obviously had an impaired taste during the summer season, as follows from the results of the second and particularly the third session. These data show that relatively large changes in taste rating of a surface water supply can occur during the year. However the results relating to the rather constant quality of the drinking water derived from ground water indicate that a different use of the taste scale by the panel members had apparently occurred during the second and the third session as compared to the first session. According to the t-test at 5% probability of exceedence, no significant differences in taste scale values existed for the drinking water derived from ground water during the second and third session.

Based on these results it was found necessary to apply a correction procedure to the taste scale values obtained during the first session of the panel, in order to be able to compare the taste assessment data of all types of tapwater studied.

The changed attitude of the panel towards the taste scale might have been caused by the fact that the panel members became more familiar with the use of the taste scale on the one hand while on the other hand data described in paragraph 3.2.2.2 also indicate that the panel as a whole was somewhat more sensitive to odorous compounds in water during the second and third session.

A correction procedure had to be applied by which the changed use of each item of the taste scale was calculated for the first session compared to the combined results of the second and third session. A detailed description of the applied correction method is given in appendix 3.2. This method results in the application of corrected scale values for the different categories on the taste scale.

By applying the original values for the taste scale categories for the data obtained during the second and third session and the corrected values for the data obtained during the first session the 20 types of drinking water can be classified as shown in table 3.9. The values relating to the two types of drinking water included in all three sessions are averages. Generally the differences in taste assessment are significant according to the t-test at 5% probability of exceedence, in case the taste scale values show a difference of 0.05 or more.

It should be noted that the part of water sample nr 4, which was presented to the panel was found to be more contaminated than the water collected in the 200 litre stainless steel vessel for chemical analysis. Therefore this water has been re-examined on the Monday, following the panel session, by a laboratory panel consisting of 12 subjects. For this reason the 95% confidence limits are less narrow for water type nr 4 than those for the other water types. The cause of this unusual finding could later on be identified as insufficient flushing of the tap before sampling, resulting in the presence of organic chemicals, released from plastic piping, in the sample contained in the 25 litre glass bottle for sensory assessment.

Table 3.9: Classification of 20 types of drinking water according to the taste rating by a panel

Tapwater type number	Main type of raw water source	Type of storage	Mean taste scale value ($\bar{x}$)	Standard deviation (σ)	95% Confidence limits 1)
1	Ground water	-	1.12	0.40	1.10 - 1.14
2	Ground water	-	1.16	0.41	1.13 - 1.19
3	Ground water	-	1.16	0.44	1.13 - 1.20
4	Ground water	-	1.16	0.56	1.00 - 1.32
5	Ground water	-	1.19	0.43	1.16 - 1.22
6	Ground water	-	1.19	0.40	1.16 - 1.22
7	Surface water	Dune infiltr.	1.20	0.43	1.17 - 1.22
8	Ground water	-	1.21	0.40	1.19 - 1.23
9	Ground water	-	1.23	0.37	1.21 - 1.25
10	Surface water	Dune infiltr.	1.30	0.32	1.28 - 1.32
11	Surface water	Dune infiltr.	1.48	0.48	1.45 - 1.51
12	Surface water	Dune infiltr.	1.55	0.44	1.53 - 1.59
13	Surface water	Dune infiltr.	1.59	0.45	1.56 - 1.62
14	Surface water	Reservoir	1.68	0.54	1.66 - 1.72
15	Surface water	Bank filtr.	1.70	0.58	1.65 - 1.75
16	Surface water	Reservoir	1.81	0.57	1.78 - 1.84
17	Surface water	Bank filtr.	1.91	0.54	1.87 - 1.95
18	Surface water	Bank filtr.	2.01	0.59	1.97 - 2.05
19	Surface water	Reservoir	2.06	0.47	2.04 - 2.10
20	Surface water	Bank filtr.	2.06	0.57	2.05 - 2.11

1) calculated according to the formula: $\mu = \bar{x} \pm 1.96 \dfrac{\sigma}{\sqrt{n}}$ in which:

$\bar{x}$ = mean taste scale value, σ = standard deviation of individual scale values
n = number of observations

It can also be concluded from these data that dune infiltration probably results in a better taste of drinking water than storage in reservoirs or the current practice of waterworks based on bank filtration of surface waters.

As the taste of water type nr 18, 19 and 20 was classified in a category approaching the *"offensive"* category this table clearly demonstrates that improved treatment of surface water and identification of those contaminants which are responsible for the adverse taste warrants further study.

3.5.3 Results of the taste quality assessment

In a similar order as given in table 3.9 the results relating to the taste quality assessment of the 20 tapwaters are presented in table 3.10. These data show that the most obvious quality aspect of ground water is the metallic taste, followed by an earthy taste quality. The drinking water, treated by dune infiltration, shows a general increase in the importance of the earthy quality of the taste, which flavour aspect is also very pronounced in two drinking waters where open storage is part of the treatment process. Possibly growth of algae or micro-organisms is related to these findings. Also the putrid taste aspect is important in the latter two types of water which points to the same possible role of growth and decay of organisms in reservoirs.

Table 3.10: Taste qualification of 20 types of drinking water by a panel

Water type number	Type of water source	Mean % of panel members which scored in a certain taste quality category				
		Chlorine-like	Earthy	Putrid	Metallic	Faint or not perceptible
1	GW	3	13	4	12	68
2	GW	5	13	3	21	58
3	GW	5	15	7	17	56
4	GW	5	12	2	12	69
5	GW	2	15	9	20	54
6	GW	3	11	3	20	63
7	DI	3	15	10	15	57
8	GW	6	8	0	17	69
9	GW	4	12	4	8	72
10	DI	6	14	4	15	61
11	DI	7	30	12	18	33
12	DI	14	21	10	17	38
13	DI	6	21	12	26	35
14	RES	21	15	6	14	44
15	BF	15	20	16	21	28
16	RES	17	28	20	17	18
17	BF	19	14	18	30	19
18	BF	36	9	12	25	18
19	RES	12	27	24	18	19
20	BF	34	17	8	19	22

GW : Ground water
BF : Bank filtrate
DI : Dune infiltration of surface water
RES : Surface water stored in a reservoir

Very striking is the high percentage of panel members which perceive a chlorine-like taste in relation to tapwater derived from water of the river Rhine by means of bank filtration. As in these cases no chlorination is applied during treatment the chlorine-like taste must have been caused by certain chemicals which originate from the contaminated river water.

3.5.4 Factors affecting the taste rating by the panel members

3.5.4.1 *Sex and age*

In agreement with the data presented in paragraph 3.2 sex was not a significant parameter in relation to the taste rating of water, while age was a factor of some importance. For the data relating to the first panel session the Pearson correlation coefficient for taste rating of the total of all 8 types of water amounted for the sex of the panel members to 0.016 ($\alpha=0.065$) and for the age of the members to -0.046 ($\alpha=0.001$). During this session, where 5 out of the 8 types of drinking water were derived from surface water, older panel members scored slightly lower values for the average taste rating for all tapwaters than the younger ones.

3.5.4.2 *Odour sensitivity*

As can be expected, subjects which are more sensitive for odorants, like isoborneol and o-dichlorobenzene, also show a better ability to discriminate good and bad water taste. This is illustrated in relation to the first session day in figure 3.5.

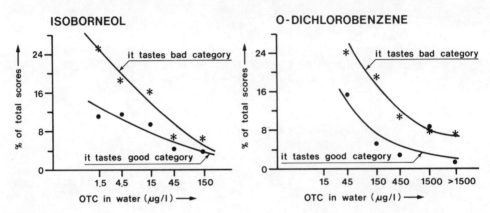

Figure 3.5
Effect of odour sensitivity of panel members for aqueous solutions of isoborneol and o-dichlorobenzene on the use of extreme taste scale categories during the first session day.

Subjects with a high odour sensitivity will generally be the first to classify a drinking water in one of the two extreme taste categories instead of the category *"it has no perceptible taste"*. The latter will be mostly used by subjects with a low sensitivity for water taste. The average difference in taste rating between water types nr 4 and nr 19 (see table 3.9) amounted to 1.95 for relatively sensitive subjects (OTC for isoborneol $\leqslant 4.5\,\mu g/l$). This difference in taste rating was only 0.66 for insensitive subjects (OTC for isoborneol $\geqslant 45\,\mu g/l$). The same effect, although somewhat less pronounced, was found in relation to odour sensitivity for o-dichlorobenzene.

These data strongly support the usefulness of a sensory sensitivity test as a means to select panel members for water tasting.

3.5.4.3 *Fatigue and habituation*

The general pattern of taste rating by the panel members during the first session day is presented in figure 3.6 for the extreme taste categories.

In the first place a striking reduction in the use of the *"it tastes good"* category occurs during the day. For the same series of 8 samples the percentage of scores in the *"it tastes good"* category decreased during the day from about 11% to approximately 3.5%. Probably part of the changed attitude to the scale, as discussed in paragraph 3.5.2 and appendix 3.2, is due to this fact. From the data relating to the scores in the *"it tastes bad"* category it is clear that an effect of fatigue or conditioning seems to results in a taste judgement which becomes increasingly negative. After the breaks the percentage of total scores in this *"bad taste"* category is restored from 12-15% to the original 7-8%.

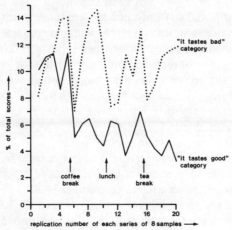

Figure 3.6
Effect of fatigue and habituation on the scores in extreme taste scale categories for the total of 8 types of drinking water studied during the first session day.

These results indicate that panel sessions aiming at comparable results should be interrupted regularly and that the session intervals should be of equal length during the different session days. The data further indicate that, preferably, intermissions should be allowed after short sessions of a panel of approximately 15 minutes. Attention should also be given to a certain familiarization of the panel members with the taste of drinking water, which will reduce the initially high number of scores in the *"it tastes good"* category.

Finally there exists a striking discrepancy between the taste assessment of the water samples during the panel sessions and the taste assessment by the panel members relating to the water at their homes, as shown in table 3.3. None of the panel members used the category *"it tastes foul"* for the tapwater supplied locally and 33% used the category *"it tastes good"*. However during the panel sessions even ground waters were scored in the category *"it tastes good"* with a frequency of only 11% of the cases or less. These data indicate the presence of a less critical attitude of consumers in relation to the taste of their local tapwater under normal conditions than in the case where a number of tapwaters is evaluated during a panel session. It is not likely that this effect should be wholly attributed to habituation. Probably the subjective presumption that the drinking water supplied locally is of good taste and wholesome to drink has also influenced the judgements.

3.6   Conclusions

Water taste is closely related to water odour, but the taste of drinking water is much better noticeable than the smell. Therefore sensory assessment of drinking water can be limited to water tasting.

According to the judgements of the panel by means of a 5-category scale as presented in table 3.6, the taste and odour of drinking water derived from ground water is considerably better than in the case of water derived from surface water. The average taste scale value for the 8 types of ground water supplies studied amounted to 1.17, while the ground water supplies showed only minor variations in taste ratings (1.12-1.23). Among the 12 surface water supplies, which showed a mean taste rating of 1.70, those 5 which are applying dune infiltration showed better taste ratings than the 7 supplies applying storage reservoirs or bank filtration (mean taste ratings were respectively 1.43 and 1.90). A particular surface water supply using a storage reservoir showed variations in taste rating values during the seasons between 1.84-2.40. This illustrated the large variations and the strongly impaired water taste which can occur. Therefore seasonal fluctuations in water taste should be studied in more detail in further investigations relating to drinking water derived from surface water.

The dominant sensory quality of drinking water derived from ground water is the metallic taste aspect and to a lesser extent the earthy taste aspect. Supplies using dune infiltration of surface water show a taste resembling to a large extent that of ground water supplies but with a more dominant contribution of the earthy taste aspect. The highest number of scores for the earthy as well as the putrid taste qualities were obtained for the drinking waters derived from Rhine water after storage in reservoirs. Growth and decay of organisms are probably the causes. Supplies receiving a high rating for the chlorine-like taste aspect where not those applying chlorination but two supplies based on bank filtration of water from the river Rhine. Residual organic compounds from the river water might be responsible for this effect.

It was found that during an inquiry panel members generally rated the taste of their local tapwater much more positive than the taste of comparable types of water presented during the panel sessions. The reason for this might be habituation but probably also a presumption that local tapwater is of good quality, resulting in a less critical attitude.

The tests on odour sensitivity for aqueous solutions of o-dichlorobenzene and isoborneol as well as the results relating to the panel sessions show a number of factors, which should be considered in the selection and application of panels for water taste assessment. In this respect the following conclusions can be derived:

1. The individual sensitivity of the sense of smell for odorous water contaminants is related to the discriminatory ability for drinking water taste. This results in a more frequent use of extreme categories of the taste scale by more odour sensitive subjects. Therefore the sensitivity of a panel can be increased by selecting subjects with a high sensitivity for one or more odorous water contaminants.
2. As about 2/3 of the panel members considered showed variations in odour sensitivity for o-dichlorobenzene and isoborneol within a factor 3, during three sessions held in a period of several months, it is recommended to measure odour sensitivity during each panel session and to exclude those subjects from a panel which show larger variations in odour sensitivity for water constituents than a factor 3.
3. As the sensitivity of the sense of smell for certain water contaminants is probably lower in case these compounds are regularly present in the drinking water, subjects used to drink tapwater derived from surface water can be less suitable as panel members than subjects used to the taste of water derived from ground water.
4. As no indications were found that sex or smoking significantly influence the odour sensitivity for o-dichlorobenzene or isoborneol it does not seem necessary to balance the number of males and females in panels or to exclude smokers.

5. As panel members show a tendency towards more negative jugdements during the course of a session it is recommended to interrupt panel sessions after relative short periods of approximately 15 minutes. For comparison of results of different session days it is essential that the session periods during each day are of the same length.

Without good instruction panels do not produce optimal results

4 Drinking water taste and inorganic constituents

4.1 Introduction

As mentioned in paragraph 1.2.3.2 drinking water should contain a minimum concentration of certain salts. To obtain a neutral taste of water the salt content should resemble the salt content of saliva to which the taste receptors are adapted (O'Mahony, 1972) (Bartoshuk, 1974). On the other hand concentrations of salts exceeding the taste threshold concentration (TTC) will result in impaired water taste (Biemond, 1940) (Bruvold et al., 1966, 1969) (Bruvold and Ongerth, 1969) (Pangborn et al., 1971) (Drost, 1971).

Only little information is available on the levels of individual salts in water at which an adverse taste becomes evident for a certain part of the population and on the relation between concentration changes and taste intensity. However such data are needed for the interpretation of the taste ratings described in Chapter 3, and in relation to recommendable levels of salts in drinking water. Therefore nine of the main salts present in drinking water have been studied with the panel described earlier.

The results of these preliminary experiments are described in this chapter and are related to the possible contribution of inorganic constituents to the taste ratings observed for the 20 types of drinking water studied.

4.2 Taste assessment of individual salt in water solutions by a selected panel

4.2.1 Stimulus presentation and test conditions

During the third panel session decribed earlier the afternoon was used for assessing the taste of individual salt in water solutions. This additional experiment was carried out under the same general conditions as described in Chapter 3 and panel members had to use the same 5-category taste scale to indicate the taste of the stimuli. The time interval between the presentation of two stimuli was 50 seconds and the oral cavity had to be rinsed with local tapwater between two presentations. The mineral composition of the tapwater used for rinsing is given in table 4.1.

Table 4.1: Mineral composition of Utrecht tapwater used for rinsing purposes

Salt ion	Concentration (mg/l)
Chloride	15
Sulphate	11
Hydrocarbonate	115
Sodium	9
Calcium	36
Magnesium	5
Iron	0.06

Nine salts detected frequently in drinking water were selected for further study: sodium chloride, sodium hydrocarbonate, sodium sulphate and the calcium and magnesium salts of these three anions. In a preliminary experiment a small panel of 12 persons determined the lowest concentration for these salts which is scored in the category *"it tastes bad"* by the majority of subjects. For each salt these concentrations were chosen as the highest concentration in a series of six to be used in the final experiment. Concentration steps with a factor 2 were used. Cups containing double-distilled water, being the solvent of all salts, were used as blanks. The salt solutions were randomized in two parts which were presented before and after teabreak. All salt concentrations were offered to each panel member only once, because of time limitations. To avoid systematic effects due to the presentation order, the sequence of the samples offered was reversed for the groups in the two adjacent rooms.

4.2.2 Results

The results of the taste assessment of different concentrations of salts in water are presented in the figures 4.1, 4.2 and 4.3. For each type of salt the value of the arithmetical average on the taste scale and the 95% confidence limits are presented. The average value for distilled water, calculated from 5 presentations to each panel member, has been used for each curve as starting point.

The results obtained indicate that addition of certain quantities of sodium hydrocarbonate or calcium sulphate to distilled water improves the taste of water. An optimal taste was found at a concentration of 1.5 meq/l of $NaHCO_3$ (125 mg/l) and at 4 meq/l of $CaSO_4$ (270 mg/l). For other salts such an optimal taste quality compared to distilled water was not clearly found.

An optimal taste quality preceeded by an impaired taste quality at lower concentrations was found for $Mg(HCO_3)_2$ and $Ca(HCO_3)_2$, while in both cases the optimal taste quality was not substantially better than the taste of distilled water. The taste of a 3 meq/l $Mg(HCO_3)_2$ solution compared to the taste of either a twice higher or lower $Mg(HCO_3)_2$ concentration was significantly better, according to the two-tailed value of t at 5% probability of exceedence. The 2.5 meq/l $Ca(HCO_3)_2$ solution tasted significantly better than the 1 meq/l solution.

A rather unchanged taste at increasing salt concentrations was found for NaCl up to a level of about 5 meq/l (290 mg/l) and for $CaCl_2$ up to a level of about 2 meq/l (120 mg/l). The taste of the 3 meq/l $MgSO_4$ solution was not significantly different from the 6 meq/l $MgSO_4$ solution.

A direct relation between taste impairment and increasing salt concentration was found for $MgCl_2$ and Na_2SO_4.

4.2.3 Discussion

According to Wipple (1907) and Lockhart *et al.* (1955) the TTC for calcium and sodium chloride, as well as for sodium, magnesium and calcium sulphate varies from 250-550 mg/l. Bruvold and Gaffey (1969) studied the taste of salt solutions at concentrations of 1000 and 2000 mg/l of salt in water. At a concentration of 1000 mg/l calcium sulphate, sodium hydrocarbonate and sodium sulphate showed the least adverse

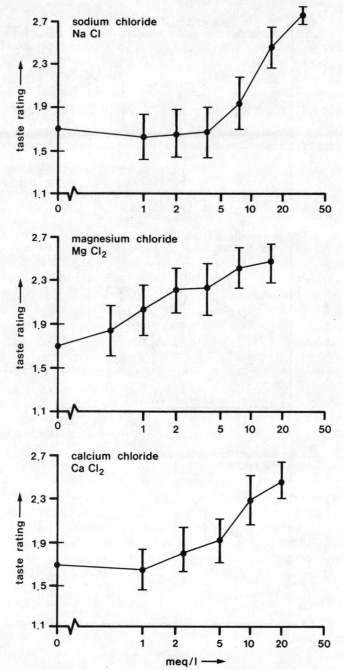

Figure 4.1
Taste assessment of aqueous solutions of sodium, magnesium and calcium chloride.

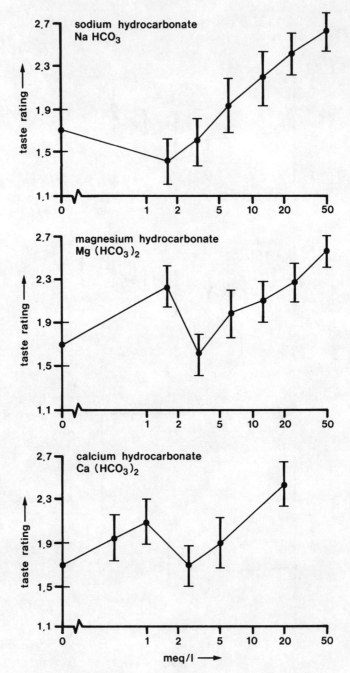

Figure 4.2
Taste assessment of aqueous solutions of sodium, magnesium and calcium hydrocarbonate.

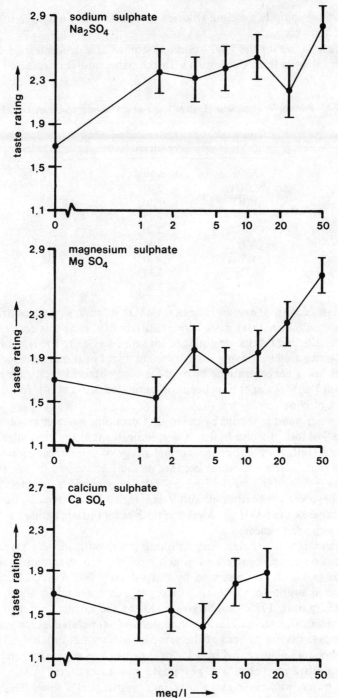

Figure 4.3
Taste assessment of aqueous solutions of sodium, magnesium and calcium sulphate.

taste while sodium chloride, calcium chloride and particularly magnesium chloride had the most objectionable taste.

Taste scale values for the 9 salt solutions studied at a concentration of 1000 mg/l have been extrapolated from the results with the panel and are presented in table 4.2.

Table 4.2: Rank order of the taste of salt solutions at salt in water concentrations of 1000 mg/l

Type of salt	Scale value (measured or extrapolated)
$CaSO_4$	1.90
$MgSO_4$	2.05
$Mg(HCO_3)_2$	2.10
$NaHCO_3$	2.15
$Ca(HCO_3)_2$	2.20
$CaCl_2$	2.40
Na_2SO_4	2.40
$MgCl_2$	2.45
NaCl	2.50

These data confirm the weak taste of a $CaSO_4$ solution at a concentration of 1000 mg/l and the mentioned strong taste of particularly $MgCl_2$ at this concentration.

Rather conflicting results are obtained for salts like NaCl. The type of scale as well as the type of water used for rinsing in both studies might be responsible for the different results. As the lower concentrations of Na_2SO_4 also obtained high ratings (see figure 4.3) solutions of Na_2SO_4 might have been contaminated in our study, for instance due to contamination of glassware.

On the other hand it should be noted that drinking water generally contains less than 1000 mg/l of total dissolved salts, which indicates that the concentrations studied by Bruvold and Gaffey are out of the normal range of salt concentration in drinking water. This is particularly important because of the fact, shown in this study, that at concentrations below 1000 mg/l different salts can show a completely different relationship between concentration and taste rating. This means that at certain concentrations below the 1000 mg/l level, the rank order of taste rating for salt solutions shows rather large differences.

Pure water is not tasteless and drinking water will only be tasteless when it contains salt ions in concentrations which are not too much different from the concentrations in saliva, as illustrated by Bartoshuk (1974) for NaCl solutions. In a parallel way consideration of the levels of inorganic components in saliva, as presented in table 4.3 (Ferguson, 1975), might give a clue to the interpretation of some of the results obtained in this study. During conditions of stimulated saliva excretion most inorganic components are present at the same levels as given in table 4.3, but sodium and hydrocarbonate are increased by a factor 2 or more in concentration. Other major constituents in saliva are potassium, phosphate and thiocyanate.

As the water used for rinsing contained substantially lower levels of sodium chloride (see table 4.1), this might have lowered the level at which the NaCl solutions started to become perceptible, but it cannot have been the cause of the rapid increase in

Table 4.3: Presence of some inorganic components in saliva (Ferguson, 1975)

Component	Concentration in mg/l	
	mean	range
Sodium	300	78 - 600
Chloride	525	
Calcium	58	24 - 110
Magnesium	10	7 - 14
Hydrocarbonate	305	

taste rating at increasing concentrations of NaCl.

The levels at which optimal taste ratings for solutions of NaCl, $CaCl_2$ and $NaHCO_3$ are given correspond reasonably to the levels of these salts in the saliva after dilution with the tapwater, used for rinsing. The results are not accurate enough to enable a detailed verification of Bartoshuk's proposed mechanism of adaptation for other salts than NaCl. An explanation for the curves obtained for $Ca(HCO_3)_2$ and $Mg(HCO_3)_2$ has not yet been found.

In a first approach to arrive at recommendations for maximum acceptable levels of salts in water to prevent an offensive taste, that concentration of the inorganic components can be considered at which the scale value of 2.07 (*"it has an offensive taste"*) is reached for 50% of the panel members. A summary of these concentrations is given in table 4.4.

Table 4.4: Concentrations of inorganic constituents in water at which the mean taste rating by the panel exceeded a value of 2.07

Constituent	Salt concentration		Concentration of the cation
	meq/l	mg/l	mg/l
NaCl	8	465	185
$MgCl_2$	1	47	12
$CaCl_2$	7	350	105
$NaHCO_3$	7.5	630	175
$Mg(HCO_3)_2$ (I)*	0.8	58	10
$Mg(HCO_3)_2$ (II)*	10	740	120
$Ca(HCO_3)_2$	7.5	610	150
Na_2SO_4	-	-	-
$MgSO_4$	14	840	170
$CaSO_4$	15	1020	300

* As the taste concentration curve for $Mg(HCO_3)_2$ twice trespasses the 2.07 value for taste rating (see figure 4.2) two concentrations are presented.

In this table the data for Na_2SO_4 have been omitted because of the possible artefact mentioned earlier. The data of table 4.4 illustrate the dominant effect on taste of the cations, and the rather equivalent effect of the cations on taste for the chlorides and the hydrocarbonates. Sodium starts to become offensive above approximately 175 mg/l, magnesium above approximately 10 mg/l, calcium above approximately 125 mg/l in the chloride and the hydrocarbonate salts, dissolved in distilled water.

This table further shows that among the three anions tested, sulphate had the weakest effect on taste. It is particularly interesting to see how the sulphate ion strongly *"suppresses"* the taste of the magnesium ion.

From the data presented in table 4.4 it may be concluded that it is more appropriate to formulate standards for cations than for anions like chloride in relation to the taste of drinking water. Based on the tentative results of these experiments with individual salts it seems that sodium generally should not be present above a level of 175 mg/l, while such a general maximum level would be 10 mg/l for magnesium and 100 mg/l for calcium.

These values, however, should be considered as tentative figures for a number of reasons. In the first place the results should be verified by studies based on a larger number of observations and including more salts as it is known that a salt like sodium carbonate has a relatively strong taste as indicated by the taste threshold which is below 100 mg/l (Lockhart, 1955). Secondly it should be noted that in practice drinking water contains a mixture of inorganic constituents, which may show additivity and masking effects. Therefore detailed further studies should also include mixtures of salts. Finally the applied taste criterion of a mean taste scale value of 2.07, corresponding with the rating *"it has an offensive taste"*, has to be refined. The presently applied criterion can be considered as the concentration at which approximately 50% of the population would rate the taste as offensive or worse, in case the same testing conditions would be applied. It seems that the maximum permissible level of salts in water should be such that only a small minority of the population exposed has to consume water of which the taste is experienced as offensive. Therefore the indicated maximum levels for sodium, magnesium and calcium might appear to be too high when the results of the recommended additional studies become available. On the other hand subjects seem to be less critical in relation to the taste of the tapwater available at their homes then in the case of panel evaluations (see paragraph 3.5.4.3).

4.3 Inorganic constituents in 20 types of drinking water and taste assessment

4.3.1 Levels of inorganic constituents in drinking water

The 20 types of drinking water, as described in Chapter 3, were analysed for the concentration of inorganic water constituents by means of current analytical methods for water quality analysis, mainly based on titrimetry, colorimetry and spectrophotometry. The levels detected for the most important parameters are presented in table 4.5, in which the same order of water types is used as in table 3.9 and 3.10. The pH of the 20 samples varied between 7.65-8.37 and the content of the carbonate ion was always below 10 mg/l. The sodium level ranged from 7 - 124 mg/l, the magnesium level from 4 - 31 mg/l and the calcium level from 29 - 173 mg/l. Iron concentrations varied from 0.01 - 0.24 mg/l.

Table 4.5: Water taste assessment and inorganic constituents in 20 types of drinking water

Water type number	Raw water source Ground water	Raw water source Surface water	Taste scale value	% of scores in "metallic" taste quality category	Conductivity	Cations				Anions		
						Fe^{3+}	Na^+	Ca^{2+}	Mg^{2+}	HCO_3^-	Cl^-	SO_4^{2-}
					μS/cm	mg/l	mg/l	mg/l	mg/l	mg/l	mg/l	mg/l
1	+		1.12	12	605	0.01	33	95	14	308	64	8
2	+		1.16	21	500	0.22	40	72	10	246	51	13
3	+		1.16	17	215	0.01	7	38	4	112	11	6
4	+		1.16	12	440	0.02	21	62	10	177	39	25
5	+		1.19	20	175	0.01	8	29	4	84	13	6
6	+		1.19	20	230	0.06	9	36	5	115	15	11
7		+	1.20	15	570	0.02	27	97	10	262	51	58
8	+		1.21	17	1140	0.04	82	173	31	634	134	1
9	+		1.23	8	190	0.06	12	24	4	66	17	13
10		+	1.30	15	875	0.18	79	114	16	270	135	78
11		+	1.48	18	965	0.01	101	104	14	182	187	98
12		+	1.55	17	840	0.01	74	100	9	195	147	72
13		+	1.59	26	675	0.05	61	85	5	182	118	86
14		+	1.68	14	635	0.08	79	54	12	104	82	108
15		+	1.70	21	895	0.02	86	104	16	274	156	30
16		+	1.81	17	1065	0.03	124	98	20	70	242	144
17		+	1.91	30	590	0.06	46	75	13	180	92	41
18		+	2.01	25	805	0.10	75	88	17	193	154	71
19		+	2.06	18	790	0.07	120	53	16	91	175	88
20		+	2.06	19	860	0.24	80	104	17	230	158	55

4.3.2 Discussion

The data given in table 4.5 indicate the absence of a strong contribution of the total of dissolved salts or a single inorganic ion to the taste of the types of water studied. Also the presence of a relation between the iron content and the metallic taste aspect cannot be concluded from these data, as water type nr 2 and 20 showed the highest iron contents, but did not receive the highest scores for the metallic taste aspect. Apparently iron does not influence water taste significantly at concentrations up to 0.2 mg/l. This means that the impaired taste of the drinking water derived from surface water must be mainly due to the presence of organic contaminants.

As far as the drinking water derived from ground water is concerned there seems to exist a small effect on water taste by the hydrocarbonate content, as shown in figure 4.4. It is of interest to note that the optimum hydrocarbonate content, which can be derived from this figure, closely approaches the hydrocarbonate content in saliva as indicated by table 4.3. The reason why the taste of water type nr 8, also a drinking water derived from ground water, deviates from the other water types might be the relatively high content of magnesium.

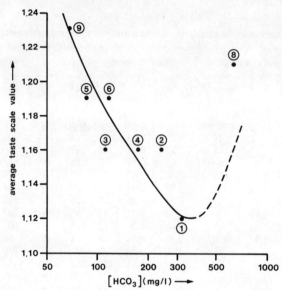

Figure 4.4
Taste rating for groundwater supplies and hydrocarbonate content of the drinking water.

These data suggest that a number of ground waters have a sub-optimal taste because they are lacking sufficient quantities of sodium hydrocarbonate.

4.4 Conclusions

Comparison of the taste assessment of 20 types of drinking water by a selected panel and the present concentrations of inorganic constituents did not show a substantial effect of these constituents on water taste. The large differences between the taste of drinking water derived from ground water as compared to drinking water derived from surface water should therefore be attributed to the presence of certain organic substances in the drinking water derived from surface water.

Among the ground water supplies the small differences in taste assessment could possibly be due to sub-optimal levels of the hydrocarbonate ion content as compared to the hydrocarbonate content of saliva of approximately 300 mg/l. In one case the relatively high level of magnesium (31 mg/l) could have resulted in a taste slightly deviating from the optimum.

Studies with the same panel, relating to the taste of increasing concentrations of 9 individual salt in water solutions, consisting of the calcium, magnesium and sodium salts of chloride, hydrocarbonate and sulphate, are in agreement with these conclusions. They further show that the cations seem to be mainly responsible for impaired taste. On the other hand the sulphate ion seems to *"suppress"* the taste of cations, compared to the chloride or hydrocarbonate ions, which effect is particularly striking for the magnesium ion. Among the three cations tested, magnesium had the strongest and sodium the weakest taste. It was further found that an optimal taste quality was obtained after addition of 1.5 meq/l (125 mg/l) of $NaHCO_3$ or 4 meq/l (270 mg/l) of $CaSO_4$ to distilled water, while addition to distilled water of the other salts had no effect or an adverse effect on taste.

The mean level at which the panel qualified the taste of the cation as offensive, in the most taste intensive salt in water solution, was tentatively found to be 10 mg/l for the magnesium ion, 100 mg/l for the calcium ion and 175 mg/l for the sodium ion. The significance of such levels for water supply practice should be evaluated by further studies.

Inorganic constituents of drinking water at the currently detected levels seem to be contributing more to the *"neutral"* or *"good"* taste of drinking water than to taste impairment. Those organic contaminants which are of potential interest to the latter problem will be discussed in the next chapter.

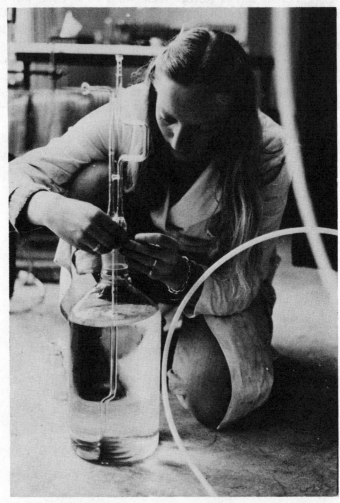

Preparation of the closed-loop gas stripping equipment for analysis of volatile compounds

5 Drinking water taste and organic constituents

5.1 Introduction

As indicated in the Chapters 1, 3 and 4 organic constituents of drinking water are probably important causes of bad taste of drinking water. In this chapter the results are presented relating to the organic compounds, detectable in the 20 types of drinking water studied by means of the analytical techniques, available at the National Institute for Water Supply (NIWS).

After a description of the analytical methods applied and the occurrence of the compounds identified, the role of individual organic compounds in taste impairment of drinking water is considered.

5.2 Analytical methods

5.2.1 General remarks

The data described in this chapter largely depend on the analytical techniques developed for analysis of trace quantities of organic compounds in drinking water. These techniques are generally described in this section. The work has mainly been carried out by G.J. Piet and his co-workers at the NIWS. Literature references are given for detailed descriptions of most of the analytical procedures applied. Those techniques which were recently developed will be briefly discussed. A general outline of the procedures, used during sampling of the 20 types of drinking water has been given in section 3.4.

Concentrates of extracts of organic compounds have been stored in micro-vials at -40°C. Analysis of the concentrates took place within 2 weeks after the moment of sampling. Blanks were analysed before and after analysis of a series of 8 samples to detect possible artefacts.

5.2.2 Organochlorine pesticides, cholinesterase inhibitors, polynuclear aromatic hydrocarbons and halogenated volatile compounds

Analysis of water samples for the presence of organochlorine pesticides and of cholinesterase inhibitors was carried out by R.C.C. Wegman at the National Institute for Public Health, Bilthoven. By means of gas chromatography (GC) and using an Electron Capture Detector, extracts were analysed for hexachlorobenzene, heptachlor (epoxide), dieldrin, α-, β- and γ-hexachlorocyclohexane, o,p'-DDT p,p'-DDT, TDE and p,p'DDE. The detection limit of this method (EX-ECD) was 0.01 microgram/litre (μg/l). The cholinesterase inhibitory activity was expressed in para-oxon units, and had a detection limit of 0.2 μg/l. At the detected levels the accuracy of both determinations was about 30% of the detected concentration.

A series of six polynuclear aromatic hydrocarbons was determined using a modified Thin-Layer Chromatographic (TLC) method of Borneff (1969), in which the solvent benzene was replaced by cyclohexane (Piet et al., 1975). The measurement was performed with a Camag TLC scanner, mounted on a Model III G.K. Turner

UV-fluorometer. Quantification of the fluorescent spots of the six compounds on the thin layer plate was accomplished by measuring the areas under the appropriate peaks of the compounds, rejusted on a strip chart recorder attached to the fluorometer. These areas were compared with calibration curves of known amounts of standard mixtures, which were run every 4 weeks. By this technique fluoranthene, 11,12-benzofluoranthene, 3,4-benzofluoranthene, 1,12-benzoperylene, 3,4-benzopyrene and 2,3-phenylene-pyrene can be detected above concentrations of 5 nanogram/litre, with an accuracy of about 100% for the low levels generally detected in drinking water.

A rapid method for the analysis of very volatile halogenated compounds in water has been used by which compounds, such as chloroform and trichloroethene, are abstracted from the equilibrated head space above a water sample. A volume of 100 ml of nitrogen gas is present above 400 ml of water sample. Equilibration takes place during 1 hour at 30°C. Subsequently a volume of 10, 100 or 1000 microlitres (μl) of the head space is injected on a UCON wide bore 5100 glass capillary column (50 m length), installed in a Tracor 550 GC, equipped with a ^{63}Ni Electron Capture Detector. The detection limit of this head space method (HS-ECD), at an injection volume of 1000 μl, is 0.01 μg/l for most halogenated compounds in water. The systematic error of the whole procedure is about 5% for an injection volume of 10 μl and at concentrations of the halogenated compounds which are about 50 times the detection limit.

5.2.3 Organic compounds concentrated by means of closed-loop gas stripping and by means of amberlite-XAD adsorption

In addition to the determination of very volatile halogenated compounds a concentration method developed by Grob *et al.* (1975, 1976) was applied. This method allows the rapid determination of a range of volatile substances with a boiling point up to 250°C. Relatively polar compounds cannot be concentrated in this way.

The concentration technique of Grob consists of a closed-loop gas stripping system by which volatile compounds are transported from the water phase to the gas phase.

The organic compounds are adsorbed from the gas phase on a small quantity of carbon. After the adsorption step the carbon filter is eluted with a few μl of carbon disulphide and this eluate is injected on a high-resolution glass capillary column, installed in a GC. As many sources of contamination may interfere, special precautions have to be taken. An air pump with stainless steel membranes was used (Metal Bellow, model MB 21, Sharon, Mass., USA). Furthermore Sovirel-Rotulex bowl-joints were applied. The glass vessel, containing the sample, and the joints with the tubing were placed under the water level to prevent diffusion of contaminants through the joints into the system. The water outside the vessel containing the sample is kept at a constant temperature of 30°. The sampled water is also used for filling the water bath. Nitrogen, the gas used for stripping, is circulated during 2 hours at a rate of 2 litres/minute through a sample volume of 5 litres. After passage through the water the stripping gas is heated to 40°C before it enters the carbon filter, to prevent condensation of water vapour on the filter, which would interfere with the adsorption process. The carbon filter, consists of 10-20 μg of carbon powder, sandwiched between two stainless steel filters in a glass tube. The filter is eluted 3 times with 8 μl of carbon disulphide. Of the combined carbon disulphide eluate 1 μl is subsequently analysed. The advantages of this method

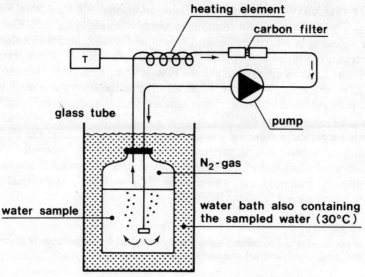

Figure 5.1
Equipment for closed-loop gas stripping of organic compounds.

are the high concentration factor of about 10^6, which can be easily obtained, and the relative small risk of introducing artefacts.

An additional method to the closed-loop gas stripping procedure was applied, consisting of adsorption of organic compounds on macroreticular resins. This method, which was developed by Junk *et al.* (1974, 1976), is more appropriate for the concentration of higher boiling and slightly polar compounds.

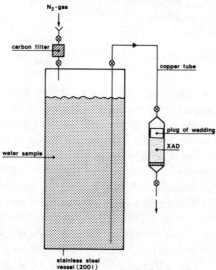

Figure 5.2
Equipment for XAD adsorption of organic compounds.

A mixture (1:1) of XAD-4 and XAD-8 was found to be most effective for a wide range of compounds. The XAD, obtained from Serva GmbH, Heidelberg, F.R. Germany, was cleaned by repeated ether extraction and was stored under methanol before use. For each water type a volume of 80 litres was pressed through a XAD column, at a rate of 50 ml/minute. The XAD column consisted of a glass tube with an inner diameter of 15 mm, filled with 25 ml of the XAD-mixture. After adsorption the resin was eluted 3 times with 30 ml or purified diethyl ether. The eluates were subsequently combined and concentrated by evaporation. For this purpose 10 ml of the eluate was concentrated by gently blowing a carefully adjusted stream of pure nitrogen gas over the liquid surface, while keeping the extract at 0°C, until a volume of 0.5 ml was reached.

Both concentration techniques have been evaluated in relation to the recoveries for a number of water contaminants. The comparable results for some typical compounds are presented in table 5.1.

Table 5.1: Recoveries after concentration of some organic water constituents by means of gas stripping and XAD adsorption (the initial concentration of each compound was 1 µg/l)

Compound	% Recovery	
	Gas stripping	XAD adsorption
Bromoform	10	90
n-Butylbenzene	90	70
n-Butylcyclohexane	100	15
Chlorobenzene	25	85
Bis (2-chloroisopropyl)ether	5	90
1, 2-Dichlorobenzene	25	85
n-Dodecane	100	10
Ethylbenzene	90	70
Hexachlorobutadiene	20	80
Tetrachloroethene	20	75
Toluene	100	60
1, 3, 5-Trimethylbenzene	95	75

The data illustrate the poor recovery of the XAD method for aliphatic hydrocarbons and the complementary characteristics of both methods.

As stated before the analytical techniques applied were not able to detect polar compounds like organic acids, lower aldehydes and lower alcolhols.

For identification of the organic compounds, the concentrates were analysed on a Finnigan 3200 gas chromatograph-mass spectrometer combination. The GC was equipped with a Grob injector and a 50 m OV 101 glass capillary column with an inner diameter of 0.3 mm. The GC column was connected by means of a glass lined tubing to the quadrupole mass spectrometer (MS). Pure helium was used as carrier gas. The GC-MS combination was coupled with an Interdata 70 computer (Wissenschaftliche Daten Verarbeitung, München, F.R. Germany). The scanning frequency of the MS was 1 scan/second in the mass range from 12-430 mass units, while the mass fragmentation patterns were produced at 70 eV ionisation energy. The spectra were compared with the 8-peak index system of the mass-spectral data centre of Aldermaston and the data stored in the Cyphernetics computer system in Ann-Arbor, Michigan, USA.

For quantification the concentrates were, in addition, analysed on a Carlo Erba 2101 GC, which was also equipped with a splitless injection system and a 50 m OV 101 glass capillary column. The compounds were detected by means of a Flame Ionization Detector (FID). Hydrogen was used as carrier gas. Generally substances present in drinking water above a level of 5 nanogram/litre could be detected in this way. The estimated concentration of each compound identified has been calculated from the response of the FID and the recovery measured for the compound, or for a closely related compound, in relation to the concentration procedure concerned. The final concentrations have been expressed on a lagarithmic basis. The concentrations may have been a factor 3 lower or higher than indicated. Concentrations of substances, approaching the detection limit of the analytical method used, have been expressed by the value of the detection limit.

Finally the odour character and odour intensity of individual compounds were evaluated by applying the *"odorogram"* technique. This technique consists of splitting the gas stream, leaving a GC-column, diverting part of the flow to a FID and the other part to the outside. Here and observer can perceive the smell of the eluting components and note his observations on the gas chromatogram. (Zoeteman and Piet, 1972/1973). In this way the results obtained with the GC-MS system have been compared with the odorograms and compounds were selected, which are possibly responsible for the impaired taste and odour of drinking water.

5.3 Detected types and levels of organic constituents in 20 types of drinking water

5.3.1 General survey

In appendix 5.1 a survey is given of the identity and the estimated concentration of organic compounds detected in 20 types of drinking water. The names have been listed according to the IUPAC system, with the exception of a number of compounds with well known trivial names, which have been described in the survey of synonyms. The data relating to the two types of drinking water, which were included in all three series of sampling, have been combined, by calculating the average values, before presentation in appendix 5.1. A total number of 280 different mass spectra has been recorded, however, 94 spectra could not be identified. Among the remaining 186 organic compounds 24 were only tentatively identified. All these 186 compounds have been included in the following general considerations.

As shown in table 5.2 the majority of organic compounds was detected in tapwater derived from surface water.

In general 39% of the organic compounds identified in drinking water were hydrocarbons containing oxygen, 27% were hydrocarbons and 15% were halogenated hydrocarbons.

Table 5.2: Occurrence of different types of organic compounds in tapwater of 8 ground water and 12 surface water supplies (see also table 3.9)

Type of compounds	% of total number of compounds identified in a tapwater category		
	Tapwater derived from ground water	Tapwater derived from surface water	All types of tapwater
C-H compounds	42	28.5	27
C-H-O compounds	39	37	39
C-H-halogen compounds	12	15	15
C-H-halogen-O compounds	0	4.5	4
C-H-N compounds	2	3	3
C-H-N-O compounds	1	3	3
C-H-S compounds	1	1	1.5
Miscellaneous compounds	3	8	7.5
Total	100	100	100
Total number of compounds	95	165	186

About 50% of the identified compounds was detected only in one or two water samples, while 10% of the compounds was found in more than 10 of the 20 types of water studied. A survey of the compounds, which were detected most frequently in the drinking water studied, is given in table 5.3.

Table 5.3: Organic compounds detected most frequently in 20 types of drinking water

Type of compounds	Name	Detection frequency*	Max. Conc. (μg/l)
Hydrocarbons	Toluene	20	0.3
	Xylenes	19	0.1
	C_3-benzenes	19	1.0
	Decanes	18	0.3
	Ethylbenzene	17	0.03
	Fluoranthene	16	0.05
	Nonanes	15	0.3
	Naphthalene	14	0.1
Oxygen compounds	Dibutyl phthalate	17	0.1
	1,1-Dimethoxyisobutane	13	0.3
	Methyl isobutyrate	13	1.0
Halogen compounds	Chloroform	16	60
	Tetrachloromethane	15	0.7

*Detection frequency: number of tapwaters, among the 20 types, in which the compound was detected

About 7% of the compounds was present in tapwater at maximum concentrations of 1 μg/l or more, while 65% of the compounds were never present above a level of 0.1 μg/l.

A list of compounds present in drinking water at concentrations above 1 µg/l is given in table 5.4. This list shows that, from a quantitative point of view, the halogenated substances and particularly the haloforms are the most important organic compounds, measured during this investigation.

Table 5.4: Organic substances detected in 20 tapwaters at levels of 1 µg/l or more

Compound	Highest detected concentration (µg/l)	Detection frequency
Chloroform	60	16
Bromodichloromethane	55	11
Dibromochloromethane	20	12
Bromoform	10	7
Trichloroethene	9	9
Trichloronitromethane	3	3
Bis(2-chloroisopropyl) ether	3	8
Octanol (isomers)	3	8
Dichloroiodomethane	1	4
1,1-Dichloroacetone	1	3
2-Cyclohexen-1-one	1	1
Methyl isobutyrate	1	13
Hexyl butyrate	1	1

5.3.2 Compounds present in drinking water derived from ground water

Among the compounds present in drinking water derived from ground water 42% belonged to the class of hydrocarbons, and about 39% belonged to the oxygen containing hydrocarbons, including alcohols, ketones, ethers and esters. Particularly striking is the presence of some simple ethers and esters, with a branched carbon chain, such as 1,1-dimethoxyisobutane, methyl isobutyrate and methyl 2-methylbutyrate. A related acid, 2-methylbutyric acid, has been reported earlier (Boorsma et al., 1969) as an accumulating product in river water after bank filtration. It is of interest to study in more detail the influence of bacteriological processes in ground water on the occurrence of these compounds and to establish their effect on water taste.

A total number of 11 halogen compounds was detected in tapwater derived from ground water. As appendix 5.1 shows, only tapwater nr 2 is completely free of detectable quantities of halogen compounds. Nowadays it seems to be an exception in The Netherlands when drinking water, which is derived from ground water, does not contain detectable levels of halogenated organic substances. Possible causes of this type of water contamination might be pollution by industrial waste discharges into the soil, contamination of the rain water which is supplying the aquifer or air contamination at the locality where aeration of the ground water, destined for human consumption, takes place. Finally it is also possible that haloforms are present in the well water due to earlier treatment of the wells with chlorine in order to restore the production capacity after clogging. A further evaluation of these possible causes is needed.

The low levels of hydrocarbons, which were measured in drinking water derived from ground water, can be of natural origin. Hydrocarbons can furthermore be introduced during treatment and transport, due to dissolution of greasing materials of pumps or due to coatings of pipes. Compounds such as dibutyl phthalate may also originate from piping materials or from air contamination.

Based on the supplies studied table 5.5 gives a general indication of the levels of some organic substances which can be expected to be present in drinking water, derived from ground water in The Netherlands.

Table 5.5: Levels of some organic constituents generally present in 8 tapwaters derived from ground water in The Netherlands

Compound	Concentration (µg/l)
Decane	0.03
Toluene	0.03
Ethylbenzene	0.03
Nonanol	0.01
Nonanal	0.03
1,1-Dimethoxyisobutane	0.1
Methyl isobutyrate	0.1
Methyl 2-methylbutyrate	0.1
Dibutyl phthalate	0.01
Chloroform	0.8
Tetrachloromethane	0.03
Tetrachloroethene	0.03

5.3.3 Compounds present in drinking water derived from surface water

Within the category of drinking water, derived from surface water, some interesting differences exist in relation to the occurrence of organic compounds. A total number of 79 organic compounds was detected in the 4 tapwaters derived from surface water by means of bank filtration, while 101 compounds were found in the 5 tapwaters prepared by means of dune infiltration and 113 compounds were detected in the 3 tapwaters, which were derived from surface water after storage in an open reservoir. In the latter type of drinking water the largest numbers of compounds containing oxygen and compounds containing halogen were found. This observation suggests that algal growth during open storage of the surface water results in the introduction of compounds containing oxygen. The high proportion of halogen compounds in water derived from reservoirs is probably due to the coincidence that these 3 supplies are the only supplies studied, which are applying break-point chorination as a part of the treatment process.

The formation of relatively large amounts of organohalides during chlorination of drinking water, containing fulvic acids and other haloform precursors, has been discussed in different studies (Rook, 1974, 1975) (Bellar *et al.*, 1974) (US EPA, 1975) (see also paragraph 1.3.4). The relationship between chlorination and production of halogenated compounds, is illustrated by table 5.6 for the 20 water supplies studied.

As a result of break-point chlorination the presence of haloform precursors such as 1,1-dichloroacetone and trichloroacetone could be demonstrated as well as the occurrence of iodine containing haloforms and trichloronitromethane. The use of lower doses of chlorine for disinfection purposes resulted in 3-10 times lower levels of the

Tabel 5.6: Effect of chlorination on the occurrence of some halogenated compounds in tapwater concentration range in µg/l)

Parameters	Type of treatment with chlorine		
	None	Disinfection	Break-point chlorination
Number of supplies studied	13	4	3
Chloroform	<0.01-2.0	<0.1 -10	25-60
Bromodichloromethane	<0.01-0.9	<0.01-10	15-55
Dibromochloromethane	<0.01-0.1	0.01- 5	3-10
Dichloroiodomethane	<0.01	<0.01-0.3	0.01-10
Bromochloroiodomethane	<0.01	<0.01-0.03	<0.01-0.3
Bromoform	<0.01	<0.01-1.0	3.0 -10
1,1-Dichloroacetone	<0.005	<0.005	0.1 -1.0
Trichloronitromethane	<0.01	<0.01-3.0	<0.01-3.0

haloforms. It goes beyond the scope of this work to discuss the potential health risks associated with the presence in drinking water of these halogenated substances (NAS, 1977).

Compounds which might be introduced into the water during open storage, as a result of growth of aquatic organisms, have been selected from appendix 5.1 on the basis of their occurrence in drinking water prepared from reservoir waters. They are summarized in table 5.7. Several of these compounds may contribute to impaired water taste.

Table 5.7: Organic compounds possibly introduced into the water during storage in open reservoirs

Compound	Max. Conc. (µg/l)	Compound	Max. Conc. (µg/l)
Octene-1	0.03	Hexanal	0.03
Nonene	0.01	Heptanal	0.1
Undecene	0.01	Octanal	0.03
4-Methyl-heptan-4-ol	1.0	Nonanal	0.1
2-Ethyl-1-hexanol	3.0	Decanal	0.1
2,6-Dimethyl-heptan-4-ol	0.03	Undecanal	0.03
trans-2-Hepten-1-ol	0.1	Heptan-3-one	0.1
trans-2-Nonen-1-ol	0.01	Dodec-2-ene-4-one	0.03
2-Methylisoborneol	0.03	2-Methylbutanenitrile	0.1
Geosmin	0.03		

The biological processes which can be involved in the production of these compounds have been generally described in paragraph 1.3.4.

Except for compounds of possible biological nature, hexachloroethane and tetrachloromethane might be present in drinking water derived from reservoir water. In this case air contamination could be the cause.

Ethyl isothiocyanate is an example of a compound probably introduced by air contamination. This substance, which is responsible for the smell of onions, is detected in water type nr 15, while this type of smell is often noticed in the neighbourhood of the treatment plant concerned.

5.3.4 Comparison between tapwaters derived from surface water and those derived from ground water

In the Chapters 2 and 3 it was shown that in general tapwater, derived from surface water, has an inferior taste compared to tapwater derived from ground water. Therefore it is of interest to consider those compounds which have been detected at substantially higher levels (for instance a factor 30 higher) in drinking waters derived from surface water than in those derived from ground water. The results of such a comparison are presented in table 5.8.

Table 5.8: Compounds detected at least at 30 times higher levels in 12 tapwaters derived from surface water than in 8 tapwaters derived from ground water

Compound name	Ratio of max. conc. for surface water versus ground water supplies	Detection freq. in surface water supplies
Nonanes	40	8
Octanols	300	6
1,1-Dimethoxypropane	> 30	3
Bis(2-methoxyethyl) ether	> 60	1
Hexyl butyrate	> 200	1
Chloroform	30	10
Bromodichloromethane	550	9
Dichloroiodomethane	> 100	4
Dibromochloromethane	2000	9
Bromoform	>1000	7
Bromochloroiodomethane	> 30	4
1,2-Dichloropropane	> 30	2
p-Dichlorobenzene	60	5
Trichlorobenzene	> 60	4
1,1-Dichloroacetone	> 200	3
Bis(2-chloroisopropyl) ether	> 600	8
Bis(dichloropropyl) ether	> 60	2
2-Chloroaniline	> 60	3
5-Chloro-o-toluidine	> 40	2
Trichloronitromethane	> 300	3
Triethyl phosphate	> 60	9
2-Chloroethyl-4-nitrophenylsulfone	> 60	1

Among the compounds listed in table 5.8 the haloforms, the chlorinated benzenes, the chlorinated ethers as well as the chlorinated anilines and toluidines could possibly contribute to impaired taste of water. The role of these compounds in this respect will be evaluated in more detail in the next paragraphs.

5.4 Organic substances and drinking water taste

5.4.1 Screening procedures to select taste impairing compounds

A first approach to elucidate the role of organic compounds in an impaired taste of drinking water, should be the selection of those compounds which can contribute individually to bad taste of water. This means that the measured concentrations should be compared with the TTC and OTC values for aqueous solutions of the compounds concerned.

Unfortunately data on taste thresholds in water are practically absent. Therefore the comparison must be limited to the data on the OTC in water. A recent compilation of OTC values by Gemert and Nettenbreijer (1977) can be used in this respect.

Only for about half of the substances identified in drinking water, OTC data have been reported. Data are lacking for the group of halogenated compounds containing oxygen, for several industrially used ethers and esters and for substituted anilines.

As it was physically impossible to determine threshold data for the nearly 100 compounds which have been detected in tapwater and for which values for the OTC in water are absent, the following procedure was applied:

1. Detected concentrations (C) of individual compounds are compared with available OTC values in water.
2. Those compounds with a C/OTC ratio of 0.01 or higher are considered as potential causes of impaired taste and odour of water.
3. The list of compounds of potential interest is compared with the lists presented in table 5.7 and table 5.8, in order to verify whether OTC data are lacking for possibly relevant substances, selected on other criteria.
4. For such compounds, detected specifically in water types with impaired taste and for which the chemical structure indicates a possibility of the presence of a low OTC, the OTC values are measured.

Within this procedure it is proposed that those substances should be considered as potential taste impairing factors, for which the C/OTC ratio is above 0.01. There are a number of reasons to consider compounds of potential interest when they are detected in water at levels which are 100 times lower than their OTC:

1. Drinking water should have a good taste, not only for the *"average citizen"*, but also for the more sensitive part of the population. Therefore low concentrations of odorous compounds which are only perceptible by such odour sensitive subjects, are still relevant for water supply practice. Earlier studies (Zoeteman and Piet, 1974) indicated that generally the most odour sensitive 4-5% of the population can detect concentrations of compounds in water at 1% of the OTC.
2. Thresholds reported in literature show large differences. The actual threshold concentrations may differ a factor 10-100 from certain values reported. Although the more recent references, excluding the most extreme values, have been used in this study it is recommendable to consider a *"safety factor"* of 10-100 for the OTC values used.
3. Most of the taste impairing compounds in drinking water will be present at sub-threshold levels. Under these circumstances it is possible that additivity and enhancement effects occur resulting in a perceptible taste of a mixture of compounds, which are not perceptible individually. Therefore sub-threshold concentrations of odorous compounds should still be considered as relevant.

4. Although threshold data are almost exclusively available for the odour of chemical compounds in water, the results described in Chapters 2 and 3 strongly suggest that tasting of water results in a more sensitive detection of the compounds than smelling. This was confirmed by orientating experiments, in which the OTC of aqueous solutions of isoborneol and p-dichlorobenzene was compared with the TTC. For taste and odour assessment 7 dilutions of these two compounds were presented twice to 10 observers. The forced choice method was used and in each presentation two of the three stimuli were blanks. The respective OTC and TTC values for isoborneol were 4.5 and 1.0 μg/l and for p-dichlorobenzene 1.0 and 0.1 μg/l. These data suggest that the TTC of an aqueous solution of a chemical may be 10 times lower than the OTC.

Based on these considerations the proposed selection procedure seems to provide a relevant first approach. On the other hand it should be realized that the presence of a compound in drinking water at concentrations below 1% of the OTC, does not quarantee that all consumers are unable to detect the taste. In a later phase more detailed studies should be carried out relating to the effect on water taste of mixtures of the compounds detected.

5.4.2 Compounds with C/OTC ratios of 0.01 or more

A survey of C/OTC ratios of taste impairing substances is given in table 5.9. The C/OTC ratios are based on the maximum concentrations measured in drinking water during this study. The highest C/OTC ratios are found for geosmin, 2-methylisoborneol, p-dichlorobenzene, decanal, chloroform and 1,3,5-trimethylbenzene. At the detected concentration the smell of 1,3,5-trimethylbenzene closely resembles the smell of Rhine water.

Table 5.9: Organic compounds probably involved in taste impairment of drinking water

Compound	Max. Conc. (μg/l)	OTC (μg/l)	C/OTC ratio
Octene-1	0.03	0.5	0.06
1, 3, 5-Trimethylbenzene	1.0	3	0.3
Naphthalene	0.1	5	0.02
Biphenyl	0.1	0.5	0.2
2-Ethyl-1-hexanol	3.0	300	0.01
2-Methylisoborneol	0.03	0.02	1.5
Geosmin	0.03	0.02	1.5
Heptanal	0.1	3.0	0.03
Octanal	0.03	0.7	0.04
Nonanal	0.1	1.0	0.1
Decanal	0.1	0.1	1.0
Heptan-3-one	0.1	7.5	0.01
Chloroform	60	100	0.6
Bromoform	10	300	0.03
Hexachlorobutadiene	0.1	6	0.02
o-Dichlorobenzene	0.1	10	0.01
p-Dichlorobenzene	0.3	0.3	1.0
1, 2, 4-Trichlorobenzene	0.3	5	0.06
β-Hexachlorocyclohexane	0.1	0.3	0.03
Bis(2-chloroisopropyl)ether	3.0	300	0.01

Several substances listed in table 5.9 are probably introduced during storage of surface water in reservoirs. Such compounds are the earthy smelling 2-methylisoborneol and geosmin, as well as octene-1 and aldehydes, like decanal.

Substances formed during chlorination, like the haloforms, may affect water taste. It can however be expected that other chemicals, with a more outspoken or unpleasant taste quality, are more relevant to impaired water taste. Chlorinated benzenes may play a role in the chlorine-like taste of waters, derived from bank filtered Rhine water. Other chemicals, not presented in table 5.9 can be also involved in this type of taste impairment.

Besides these compounds which probably impair the taste of water other substances are present which may influence the taste in a positive way. Examples of such compounds are several esters such as methyl isobutyrate and methyl 2-methylbutyrate. However reliable OTC values for these compounds are not yet available.

5.4.3 Some compounds lacking OTC references

Among the compounds presented in table 5.8 some substances are of potential interest in relation to the chlorine-like taste of the bank filtered type of tapwaters. The chlorinated compounds particularly present in bank filtered waters are 2-chloroaniline and 5-chloro-o-toluidine.

OTC estimates for solutions of these compounds in water were carried out, using the forced choice method and presentations of three stimuli of which two were blanks. A total number of 5 dilutions of these compounds was presented twice to 15 subjects. The estimated OTC values of these substances and some related compounds are given in table 5.10. The C/OTC ratios for 2-chloroaniline and 5-chloro-o-toluidine are respectively 0.1 and 0.06, which shows that both compounds are probably involved in taste impairment. Furthermore the taste and odour quality of these compounds resembled the flavour of drinking water derived from bank filtered Rhine water.

Table 5.10: Estimated OTC values of aqueous solutions of some chlorinated organic compounds

Compound	OTC ($\mu g/l$)
2-Chloroaniline	3
3-Chloroaniline	100
4-Chloroaniline	10
3,4-Dichloroaniline	3
5-Chloro-o-toluidine	5
1,2,4-Trichlorobenzene	5

Besides monochloroaniline and 5-chloro-o-toluidine, 3,4-dichloroaniline also seems to be of potential significance. This compound was detected in water type nr 17, at a level of $0.03 \mu g/l$, which means that the C/OTC ratio amounted to 0.01. The estimated OTC value for 1,2,4-trichlorobenzene of 5 $\mu g/l$ is the same as the OTC value reported by Kölle et al. (1972) for this compound.

By means of the odorograms a number of strongly smelling compounds have been detected which could not yet be identified. Therefore the proposed causes of impaired water taste should be considered more as a starting point than as a final conclusion.

It is recommended to collect further OTC data based on large groups of subjects for those organic compounds which are detected in drinking water and for which OTC data are still lacking. Such an exercise will certainly result in the identification of other taste impairing compounds.

5.5 Conclusions

A total number of 280 organic compounds has been detected in the 20 types of drinking water investigated. Nearly 100 compounds of these 280 could not be identified and 24 substances were only tentatively identified by means of mass spectrometry.

Although the analytical techniques applied were not suitable to detect polar organic compounds, the category of oxygenated hydrocarbons, containing 39% of the total number of identified substances, turned out to be the largest. Of the compounds identified 27% belonged to the group of hydrocarbons and 15% belonged to the halogenated hydrocarbons.

Several alkanes, lower alkylated benzenes and phthalates were detected in practically all types of drinking water. From a quantitative point of view the haloforms were most important.

In ground water supplies some halogenated hydrocarbons were rather frequently detected. Examples in this respect are tetrachloromethane, trichloroethene and tetrachloroethene, which were probably present in the drinking water as a result of ground water contamination. Furthermore ground water supplies often contained esters, like methyl isobutyrate and methyl 2-methylbutyrate, in concentrations of 0.1-$1.0\,\mu g/l$. These compounds might contribute to the pleasant taste of certain ground waters. The latter should be verified by TTC determinations for aqueous solutions of these compounds.

Nearly twice as much organic compounds were found in surface water supplies than in ground water supplies. The highest number of organic compounds was found in tapwater derived from Rhine water after storage in open reservoirs. This was probably due to the combined effect of the formation of organic compounds by organisms, such as streptomycetes and blue-green algae, and by break-point chlorination. In these drinking waters potential taste impairing compounds of biological nature are the earthy smelling geosmin and 2-methylisoborneol, octene-1, 2-ethylhexanol and several aldehydes, like nonanal and decanal.

It was found that water treatment by increasing degrees of chlorination resulted in increasing concentrations of haloforms, up to the $100\,\mu g/l$ level for total haloforms. Also the formation of concentrations up till the $\mu g/l$ level of 1,1-dichloroacetone, trichloroacetone and trichloronitromethane was found under such circumstances. Although OTC values are not yet available for each of these compounds, it is anticipated that due to the rather sweet odour of haloforms, chlorination does not contribute substantially to taste impairment of drinking water in the cases studied.

The reported chlorine-like taste of drinking water prepared from Rhine water after bank filtration could be related to a number of chlorinated compounds. The following chlorine containing substances were present in tapwater at concentrations above 1% of their OTC:

- o- and p-Dichlorobenzene
- 1,2,4-Trichlorobenzene,
- Hexachlorobutadiene,
- Bis (2-chloroisopropyl) ether,
- 2-Chloroaniline,
- 3,4-Dichloroaniline,
- 5-Chloro-o-toluidine.

Odorograms have shown that a number of strongly smelling compounds present in drinking water have not been identified in this study. Furthermore OTC and TTC data for many compounds identified in drinking water, are still lacking. It is recommended to continue the work described in this chapter, as the results obtained are only a beginning of the identification of taste impairing compounds in drinking water. Through future investigations many other compounds most probably can be added to the list presented.

Finally it will become necessary to investigate interactions of mixtures of the compounds identified at the detected levels.

Division for water quality analysis by gas chromatography techniques

6 Sensory assessment of drinking water quality and health protection aspects

6.1 Introduction

According to the WHO International Standards for Drinking Water (1971), *"water intended for human consumption must be free from organisms and from concentrations of chemical substances that may be a hazard to health. In addition, supplies of drinking water should be as pleasant to drink as circumstances permit".*
This statement seems to imply that sensory assessment of water contamination is of less importance than contamination which may cause disease. Although this is certainly not the official policy of WHO it is a philosophy which is often presented in discussions relating to standards for taste and odour of drinking water. It should be noted that impaired taste is on its own a health relevant aspect of water quality. Furthermore it is of interest to know to which extent taste and odour are indicating the presence of water contaminants, directly related to human diseases. It is the purpose of this chapter to investigate the presence or absence of a relationship between perceptual and toxicological effects of drinking water.

As discussed in section 1.1 sensory quality assessment has been considered by several authors as a useful means by which man is warned in case hazardous chemicals are present in his environment. Summer (1970) even supposed a very close relationship between sensory detectability of natural compounds and health protection:
"Der Mensch wird die Flucht ergreifen oder sonst eine zu seiner Rettung beabsichtigte Handlung ausführen, wenn er ein ihm gefährliches Gas riechen kann. Dieses Warnsystem hat sich im Verlauf der Entwicklung herausgebildet und sichert das Leben des Tieres oder Menschen".

In order to verify the supposed warning function of the chemical senses, the effect of sensory water quality assessment on water consumption habits, as detected by the inquiry described in Chapter 2, will be considered in more detail. In case the warning mechanism actually functions it may be expected that a reduced water consumption is found for those individuals supplied with water of bad taste and smell.

A second aspect, which needs to be verified, is the effectiveness of the warning mechanism of chemoreception. The most optimal criterion for this purpose would be a comparison of the Odour Threshold Concentration and the Maximum Acceptable Concentration of chemicals in water in view of possible chronic toxic effects or certain subacute effects such as changes in light sensitivity of the eye due to sensory detection of air pollutants (Ryazanov, 1962). Presently available toxicological data are the Maximum Acceptable Daily Intake (ADI) values which have been recommended for pesticide residues in food (WHO, 1973, 1976). After a comparison based on such ADI-values, which only relate to a limited number of compounds, the effectiveness of chemoreception as a warning system for acute lethal effects, will be considered. Although this exercise is of less direct value for water supply practice, as it is practically impossible to find acute lethal levels of chemicals in drinking water, it will enable the verification of the hypothesis of Richter (1950) and Summer (1970) that the chemical senses will only fail as a sensitive warning instrument for toxic compounds, in case these substances are of man-made origin.

Finally some consideration will be given to the possibility that the presence of known or suspect carcinogens for men or animals is accompanied in practice by the presence of other compounds in water which are detected more easily by the human senses.

6.2 Sensory water quality assessment and water consumption

Based on the data obtained from the inquiry, as described in Chapter 2, a general survey of the effect of the assessment of taste, odour, colour and temperature on the quantity of water consumed daily is given in table 6.1. According to these data only taste and odour are affecting total water consumption. Generally a quantity of 1.26 litre/head/day of tapwater is consumed by adults in The Netherlands. Of this quantity 81% is consumed at home, 25% is consumed as tea, 18% is consumed as pure water and 4% is consumed as lemonade.

The differences in the total quantity of tapwater consumed can be completely attributed to the reduced consumption of drinking water as such, which amounted to 0.32 l/h/d for those assessing taste as good and only to 0.15 l/h/d for those assessing taste as offensive or worse. Furthermore it is amazing that the quantity of tea consumed is hardly affected by the taste and odour of the drinking water. This was confirmed by the data on the actual flavour of the tea. These data showed no reduction in tea consumption for those assessing the flavour of the tea as unpleasant.

Table 6.1: Water consumption and sensory water quality assessment (n = 3073, 18 years and over)

Sensory aspect	Category	n	Average quantity of tapwater consumed daily (litre/head)			
			Total	Total at home	Total as drinking water	Total as tea
Taste	Good	927	1.36	1.12	0.32	0.32
	Not perceptible	1615	1.23	1.00	0.20	0.31
	Faint, hardly offensive	320	1.19	0.92	0.20	0.30
	Offensive or worse	211	1.17	0.97	0.15	0.31
Odour	Good	234	1.46	1.28	0.32	0.36
	Not perceptible	2583	1.25	1.01	0.23	0.30
	Faint, hardly offensive	158	1.17	0.93	0.18	0.31
	Offensive or worse	98	1.23	1.04	0.14	0.39
Colour	Colourness, clear	2579	1.26	1.02	0.24	0.31
	Sometimes slightly coloured	392	1.28	1.07	0.21	0.31
	Mostly not offensive	61	1.15	0.97	0.16	0.28
	Mostly offensive or worse	41	1.24	0.93	0.24	0.31
Temperature	Fresh	2415	1.24	1.01	0.23	0.31
	Somewhat tepid or worse	658	1.31	1.06	0.24	0.31
Total		3073	1.26	1.02	0.23	0.31

After regrouping of the data for 50 communities, as described in appendix 3.1, calculations of linear regressions showed that water taste is indeed a major factor, among the variables considered, which determines the quantity of drinking water consumed as such. The coefficients of correlation obtained are presented in table 6.2. Besides water taste the judgement on *"water taste elsewhere"* considerably affects water consumption.

Table 6.2: Coefficients of correlation for the daily consumed quantity of drinking water as such and water quality assessment aspects

Water quality assessment aspects	Coefficient of correlation (r)*
Taste elsewhere rating	-0.51
Taste rating	-0.47
Acceptability rating	-0.44
Colour rating	-0.33
Safety rating	-0.26

*$|r| \geq 0.28$ for $\alpha \leq 0.05$

The highest coefficient of correlation ($r = 0.53$) was obtained for the combination of taste rating and the judgement on water taste elsewhere:

Water consumption (l/h/d) = 0.31 - 0.04∗ taste rating - 0.11∗ taste elsewhere rating

A remarkable conclusion which may be drawn from the data of table 6.2 is that a correlation between the safety rating and water consumption, which was originally postulated, could not be confirmed. These data indicate that water taste assessment is even more important for the actually consumed quantity of drinking water than acceptability rating or safety rating. In this respect chemoreception indeed seems to function as a warning mechanism.

In the case of tea consumption, which is a less easily changed part of the daily consumption pattern, even an unpleasant taste does not affect the quantity consumed. In the case of coffee and soup the bad taste is masked and does not influence the consumption pattern at all. Therefore perceptual water quality affects the total quantity of tapwater consumed only to a minor extent. This means that the warning signals received by the chemical senses are not translated systemically by the consumer into a need to avoid the use of such bad tasting water for consumption purposes. This is only done for the minor quantity of water, which is consumed without any additional treatment in the household, and for which many substitutes, although about 1000 times more expensive, are readily available. An additional reason for this suppression of signals of the sensory warning system is the difficulty in obtaining an alternative supply of drinking water for general use at home.

Only in extreme cases of bad water taste the consumer will collect his water by means of buckets and tanks from other sources. Such a practice might result in increased health risks from a microbiological point of view.

6.3 Sensory detectability and chronic toxic effects of compounds

In this section the problem will be discussed whether water, which is free of taste and odour, is safe to drink and whether water which has an impaired flavour should be considered to be a potential danger to health. For this purpose a comparison has to be made of OTC values of chemical compounds in water and the concentrations corresponding with the acceptable daily intake (ADI) of these chemicals.

The ADI of a chemical is defined as the daily intake which, during an entire lifetime, appears to be without appreciable risk on the basis of all the known facts at the time (WHO, 1976). The ratio between the OTC and the ADI is considered critical in case an adult person, with a body weight of 70 kg, in 24 hours consumes 2 litres of drinking water containing a concentration of the chemical corresponding with the OTC (g/m^3), while the person is exposed in this way to the ADI (mg/kg body weight/day). This critical ratio amounts to 35 ($g.m^{-3}.mg^{-1}.kg$.). The critical ratio is of course only an approximation as the OTC applies to 50% of the individuals while the ADI applies to 95-99% of the individuals in a population. Calculated OTC/ADI ratios for 12 pesticides are presented in table 6.3. The data are based on OTC values compiled by Gemert and Nettenbreijer (1977) and on ADI values as recommended by the Joint FAO/WHO Meetings in 1972 and 1975 on Pesticide Residues in Food (WHO, 1973, 1976).

These ratios are clearly below the critical value for 3 compounds, 6 have about the same order of magnitude, whilst for aldrin, dieldrin and γ-HCH the OTC/ADI ratios clearly exceed the critical value. So for at least 3 out of the 12 pesticides considered the chemical senses would not warn the consumer against the presence of concentrations of these toxic compounds which exceed the ADI values. In these cases the consumer might however be warned by the presence of accompanying odour intensive compounds. This aspect will be discussed in more detail in section 6.5.

Table 6.3: OTC and Maximum Acceptable Daily Intake (ADI) values for some pesticides

Compound name	OTC (g/m^3)	ADI (mg/kg body weight/day)	OTC/ADI ratio
Aldrin	0.017	0.0001	170
sec-Butylamine	5*	0.2	25
Chlordane	0.0005	0.001	0.5
2,4-D	3.1	0.3	10
DDT	0.35	0.005	70
Dieldrin	0.041	0.0001	410
Endrin	0.018	0.0002	90
γ-HCH	12	0.01	1200
Heptachlor	0.02	0.0005	40
Malathion	1.0	0.02	50
Methoxychlor	4.7	0.1	47
Parathion	0.04	0.005	8

*estimate on basis of related amines

6.4 Sensory detectability and acute toxic effects of organic compounds

6.4.1 General aspects

According to Winneke and Kastka (1976, 1977) odour nuisance due to air pollution is a 3-dimensional phenomenon, consisting of sensory perception, socio-emotional reactions and somatic aspects, including induction of vomitting and interference with normal breathing. In the case of perceptible contamination of drinking water such acute physiological effects are less likely to occur, mainly because of the reduced levels of contamination present after water treatment and because of the possibility of the consumer to avoid contact with the water.

Furthermore from a theoretical point of view it is unlikely that compounds in water are present in acute toxic concentrations before they can be smelt. This can be illustrated on the basis of OTC data of chemicals in water in comparison with the lethal dose in 24 hours of the chemicals for 50% of experimental animals (LD_{50}). These data are presented in the appendices 6.1 and 6.2. The OTC values are derived from the compilation of Gemert and Nettenbreijer (1977) and the LD_{50} data are taken from "*The Toxic Substance List, 1974*" of Christensen, *et al.* (1974).

Generally LD_{50} values are based on orally administered doses to the rat. A critical ratio for acute lethal effects can be defined as the OTC/LD_{50} value at which the LD_{50} is reached within 24 hours for an adult person of 70 kg body weight after drinking 2 litres of water, containing the chemical at a level equal to the reported OTC. This critical ratio for acute lethal effects amounts to 35 ($g.m^{-3}.mg^{-1}.kg$) for water contaminants.

Table 6.4: Frequency distribution of 93 chemical contaminants in water over different OTC/LD_{50} ratio classes (Critical Ratio (CR) is 35)

% of total number of compounds						
$<10^{-5}$CR	$(10^{-5}\text{-}10^{-4})$CR	$(10^{-4}\text{-}10^{-3})$CR	$(10^{-3}\text{-}10^{-2})$CR	$(10^{-2}\text{-}10^{-1})$CR	$(10^{-1}\text{-}10^{0})$CR	$\geq$CR
43	30	16	10	1	0	0

As table 6.4 shows no examples of water contaminants are reported for which human consumption within 24 hours of 2 litres of water, containing the OTC of the chemical, constitutes a lethal dose or 10% of the lethal dose, in case the lethal dose for men is supposed to be equal to the dose observed for experimental animals. Water contaminants with the highest values for the OTC/LD_{50} ratio are formic acid, ethanol, acrylonitrile, aniline and lindane. Fortunately the presence of these compounds is often accompanied by other more odorous substances. The lowest ratios are found among natural flavour compounds such as β-ionone and decanal.

6.4.2 Natural and industrial compounds

The listed LD_{50} values and OTC/LD_{50} ratios furthermore provide an interesting base to verify the hypothesis of Richter (1950) and Summer (1970) that the chemical senses are detecting toxic levels of naturally occurring substances more effectively than toxic levels of compounds introduced by industrial discharges.

Using the compiled OTC/LD_{50} data of appendices 6.1 and 6.2 as an approximate indication of the effectiveness of the sensory warning system for detection of acute toxic concentrations of chemicals in water, a comparison for natural and industrial compounds can be obtained. Such a comparison has to be rather arbitrary. Compounds formed biologically but distributed on earth mainly by industrial activities, like a number of oil compounds, have been considered as industrial compounds. Compounds such as aliphatic alcohols, aldehydes and acids, although also produced industrially, have been considered as natural compounds due to their general presence in vegetables, body fluids, etc..

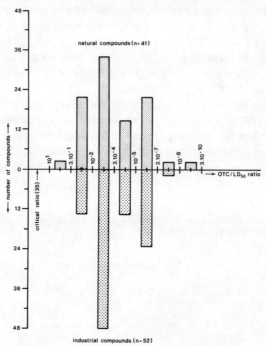

Figure 6.1
Frequency distribution of OTC/LD_{50} ratios for 41 natural and 52 industrial compounds.

From the results presented in figure 6.1, it is clear that so-called industrial compounds are detected in the same way and with a comparable sensitivity as the natural compounds. Among the compounds studied the industrial compounds tend to show lower OTC/LD_{50} ratios than the natural compounds. It may even be stated that the sense of smell has a remarkable sensitivity for a number of man-made compounds to which it has probably never been exposed before in evolutionary history. Typical examples in this respect are chlordane, chlorophenol, dichloro- and trichlorobenzene.

From the data compiled in the appendices 6.1 and 6.2 it appears that the chemical senses generally act as very sensitive sensors which are non-specific from a toxicological point of view, but which in many cases give a detectable signal at concentrations of compounds which are considerably lower than the levels resulting in acute lethal effects. In exceptional cases there may exist a relation between taste threshold and biological activity as suggested for substances like *l*- and *d*-quinine, *l*- and *d*-amphetamine and a few other structurally specific drugs (Fisher and Griffin, 1964).

6.5 Carcinogens and accompanying odorous compounds

A number of compounds like certain pesticides may be present in drinking water above the level corresponding with the maximum acceptable daily intake, without being perceived by the chemical senses. Therefore it is of great interest to investigate whether or not such contaminated water is in practice also contaminated by a number of odorous compounds. Such an indicator function of odorous compounds is not unlikely. Data described in previous chapters showed that drinking water types containing the highest amounts of organic chemicals were also those with the most objectionable taste. In this respect it would be relevant to compare the sensory water quality aspects with the presence of known or suspected carcinogenic substances like certain haloforms, tetrachloromethane and 3,4-benzopyrene (US EPA, 1975, 1977). These are generally not present in concentrations detectable by the sense of taste or smell. A possible adverse effect on health of the low concentrations of such carcinogens in drinking water has been suggested by epidemiological studies in Louisiana in which increased death rates due to several types of cancer were found in communities supplied with water from the contaminated Mississippi river (Page *et al.*, 1976).

In order to answer the question whether the sense of smell is indirectly detecting water contamination by carcinogens, the chemical data described in Chapter 5 and the taste assessment of the 20 tapwater types by the panel, as described in Chapter 3, were compiled. From these data a number of chemicals classified as known or suspected carcinogens, bacterial mutagens or teratogens (Kraybill, 1976, 1978) (US EPA, 1977) were selected as shown in table 6.5. For those compounds listed as animal carcinogens it is of course not at all certain that they could be carcinogenic for man. Table 6.5 confirms the hypothesis that for several compounds like tetrachloromethane, the haloforms and bis(2-chloroethyl) ether, increased concentrations correspond with increased taste impairment. However for other toxic chemicals like trichloroethene and 3,4-benzopyrene such a relationship is absent or even reversed. The reason is that accidental ground water contamination, which caused the observed increased levels of trichloroethene, or release of polynuclear aromatic hydrocarbons or phthalates from piping materials, is generally not accompanied by increased levels of other compounds which are more easily perceptible by the senses.

The data suggest, that there may be a relationship between taste assessment and the presence of suspected carcinogenic, mutagenic or teratogenic compounds in drinking water, in case these compounds are accompanied by a wide range of organic substances. This will generally be the case when heavily contaminated surface water is used as a raw water source. In those situations where a single compound or a single source of contamination is reponsible for the presence of suspected carcinogenic chemicals in the drinking water, it is highly unlikely that the increased level of such toxic chemicals is accompanied by a bad taste and smell of the water. Even in an exceptional situation where such compounds themselves could be smelt or tasted, as can be the case with chloroform and trichloroethene, their weak and sweet taste will not easily alarm the consumer.

Therefore the absence of any adverse taste or smell of drinking water certainly does not guarantee that the drinking water does not contain quantities of carcinogens or teratogens which could create a potential health hazard.

Table 6.5: Presence of some known or suspected carcinogens, bacterial mutagens and teratogens in drinking water, belonging to different taste categories.

Compound and toxicity category	Concentrations (microgram/litre) detected in drinking water* of a certain taste category**		
	1.10 - 1.45	1.45 - 1.80	1.80 - 2.15
Suspected human carcinogens			
- Benzene	<0.01 - 0.03	<0.01 - 0.1	<0.01 - 0.1
- 3,4-Benzopyrene	<0.005 - 0.01	<0.005 - 0.1	<0.005 - 0.005
Animal carcinogens			
- Lindane	<0.01 - 0.01	<0.01 - 0.1	<0.01 - 0.02
- Chloroform	<0.01 - 2.0	<0.01 - 25	0.2 - 60
- Tetrachloromethane	<0.01 - 0.1	<0.01 - 0.1	<0.01 - 0.7
- Trichloroethene	<0.01 - 9.0	<0.01 - 3.0	<0.01 - 0.5
Suspected animal carcinogens			
- Bis (2-chloroethyl) ether	<0.005	<0.005	<0.005 0.03
Bacterial mutagens			
- Dibromochloromethane	<0.01 - 0.3	<0.01 - 5.0	<0.01 - 20
- Bromoform	<0.01 - 0.3	<0.01 - 3.0	<0.01 - 10
Teratogens			
- Nicotine	<0.005	<0.005 - 0.005	<0.005 - 0.03
- Diethyl phthalate	<0.005 - 0.03	<0.005 - 0.1	<0.005 - 0.03
- Dibutyl phthalate	<0.005 - 0.1	0.01 - 0.3	0.005 - 0.1

* Classification of water types given in table 3.9.
** Definition of taste scale values given in table 3.6.

6.6 Conclusions

The results of the inquiry carried out among 3073 adults in The Netherlands in 1976 have shown that sensory water quality assessment may to a certain extent function as a warning mechanism for hazardous water contamination. Adults assessing water taste as offensive or foul consumed only 50% of the drinking water as such, as compared with those assessing water taste as good. Sensory water quality assessment was even more important for the quantity of drinking water consumed daily than the reliability rating of the water ($r = 0.47$ for water taste rating and drinking water consumption and $r = 0.26$ for reliability rating and drinking water consumption).

The warning signals generated by the chemical senses do not result in a reduced consumption of fluids derived from tapwater such as tea, coffee or soup. This phenomenon is most striking for the unaffected consumption of tea even when the tea has a bad taste. The background of this finding might be that the consumption pattern is less easily adapted to the water quality in the case of tea than in the case of drinking water, for which many suitable substitutes are available in the households. The same reasoning applies to the consumption of coffee and soup, in which the masking of impaired water taste and odour further suppresses the sensory warning signals.

Because of this attitude of the consumer the sensory warning system only affects to a minor extent the actually consumed quantity of tapwater, which amounts to 1.36 l/h/d for those assessing water taste as good and to 1.17 l/h/d for those perceiving an offensive or foul taste of water. This means that chemoreception of water contaminants only functions in a rudimentary and limited way as a mechanism in health protection.

Furthermore the detection threshold for hazardous chemicals, exerting long term toxic effects, is not sufficiently low and specific to prevent exposure to levels exceeding the maximum acceptable daily intake for certain compounds such as aldrin, dieldrin and lindane. Also a number of compounds, recognized as animal carcinogens such as chloroform and trichloroethene will not be perceived at concentrations in water which possibly increase the risk of cancer. In this respect the sensory warning system is not sufficiently effective as a means of health protection.

On the other hand no chemicals are presently known for which consumption by adults within 24 hours of 2 litres of water, containing the chemical in a concentration below the OTC, would probably result in acute death.

Contrary to statements by Richter (1950) and Summer (1970), the warning function of the sense of smell for acute lethal effects seems to be as effective for natural compounds as for industrial compounds. This suggests a non-specific sensitivity of the chemical senses in relation to chemicals having different toxic effects.

It was found that a useful indirect sensory warning for the presence of suspected carcinogens is obtained only in the case where such toxic substances originate from sources, contaminated by a wide range of substances. In those cases where a single suspected carcinogen or a single source is responsible for the contamination, the sensory warning system will probably fail.

Therefore the absence of any adverse taste or smell does not guarantee the safety of the drinking water, although on the other hand impaired water taste often may coincide with the presence of compounds possibly exerting carcinogenic or other chronic toxic effects.

Modern laboratories accentuate advanced analytical instrumentation and forget the usefulness of sensory techniques

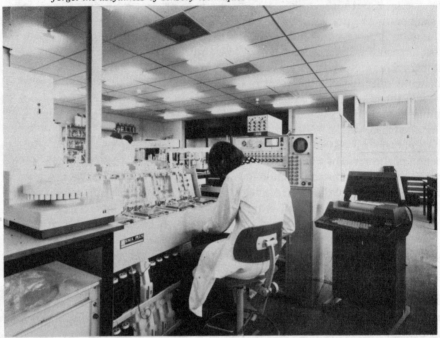

Automatic analysis of inorganic components in water samples

A GC - MS computer set up for the analysis of trace organics in water

7 Further study and application of water quality assessment

7.1 Introduction

The practical consequences of the combined results presented in the preceeding chapters are evaluated in more detail in this final chapter.

For this purpose first the value of sensory water quality assessment will be considered followed by possible applications of the determination of sensory aspects of water quality. Subsequently the main causes of the impaired taste and odour of water are considered and measures necessary to reduce perceptible contamination of drinking water are discussed. In the final paragraph the need for further studies and the kind of problems for which water quality assessment can be most successfully applied are considered.

7.2 Value of sensory water quality assessment

In Chapter 6 chemoreception of water contaminants has been generally shown to provide on the one hand a useful mechanism to warn the consumer of drinking water against the presence of contaminants in acute lethal concentrations. On the other hand the absence of any adverse taste or smell does not guarantee the safety of the drinking water for consumption during a lifetime. It was illustrated that impaired water taste often coincides with the presence of compounds possibly exerting carcinogenic or other chronic toxic effects. Therefore the generation of a signal by the chemical senses is a phenomenon to be taken seriously both by the consumer and by those responsible for the production and control of drinking water.

It was found, however, as described in section 6.2, that the *"warning mechanism"* of chemoreception functions only in a rudimentary and limited way as far as the consumer is concerned. This will be partially caused by the absence of a suitable alternative source of drinking water from which the consumer can prepare his tea, coffee and soup.

The taste and odour of drinking water derived from surface water were rated as considerably impaired in contrast with taste and odour of water from ground water supplies. This finding, described in the Chapters 2 and 3, shows two important facts. Firstly, sensory assessment of water quality provides a useful overall indicator for chemical contamination of drinking water (see also table 5.2). Secondly, drinking water derived from surface water is generally given an insufficient treatment to prevent the perception of an adverse taste and smell of the water by a relatively large group of the consumers. This means that perceptible contamination of drinking water seems to be underestimated by those responsible for the preparation and control of drinking water, derived from surface water.

It is recommended to improve this situation by applying statistically justifiable sensory methods and standards for drinking waters, in combination with more stringent water treatment procedures. Furthermore better regulations are needed in relation to waste water discharges into surface waters which are also used for the production of drinking water.

7.3 Application of sensory water quality assessment

7.3.1 Relation between type of water and method of taste or odour intensity determination

Taste and odour intensity measurements are used to control waste water discharges, as a check on the quality of surface water for various uses, for the determination of treatment dosages and as a test of the effectiveness of different kinds of individual treatment steps and of the water supply system as a whole, and as a means of tracing the source of water contamination. Waste waters can show odour numbers (ON) of 100-10.000 or more, while contaminated river water may have ON values varying from 10-100. Drinking water is often tasteless and odourless for a large part of the consumers. For this reason different methods have to be applied for different types of water. This can sometimes lead to conflicting situations when the taste or odour intensity of water has to be followed from a heavily contaminated raw water through subsequent stages of treatment up till the point where the treated drinking water is delivered to the consumer (see also paras. 1.2.2 and 1.2.3). As long as the water has an easily observable smell (ON above 3-10) the determination of the odour intensity can be preferred over the more sensitive but also more complicated determination of the taste intensity. An additional factor is that observers should not taste waters which are bacteriologically contaminated, as is practically always the case with waste water and surface water. Some laboratories try to overcome this problem by chlorinating the sample and by subsequently removing the excess chlorine by adding thiosulfate. Due to the probable formation of chlorinated more or less odour intensive products this procedure has to be discouraged.

In the case of waters with low odour intensities, and which are relatively safe from a bacteriological point of view, determination of the taste intensity has to be preferred. As relative large panels and laborous methods are needed for direct odour and taste intensity measurements, procedures have been proposed which allow for a more rapid indirect estimation of the taste or odour intensity. Examples in this respect will be discussed in more detail for the case of contaminated waters on the one hand and drinking waters on the other.

7.3.2 Determination of odour intensity of contaminated waters

7.3.2.1 *General aspects of the Odour Number determination*

A number of useful instructions for a widely used odour intensity determination by means of the method of limits is given in the Standard Methods (Rand *et al.*,

1975). According to the Standard Methods the odour intensity is defined as that dilution of the sample at which the odour is just perceptible. The measurement of the Threshold Odour Number (TON) obtained in this way has, however, some disadvantages, although the determinations can be rapidly carried out.

It must be realized that the accuracy of the results is rather poor. Samples are presented only once to the observers and the danger exists that panel members start to guess that samples are odour free or not in the sequential presentation of dilutions from low to high concentrations. Also when the dilutions are presented in a randomized way guessing will play a role. As stated in paragraph 1.2.2.2, the forced choice method should be preferred to overcome guessing and large influences of contaminants in the dilution water. According to this method dilutions together with one or more blanks are presented in a randomized order. A further advantage of the forced choice method is that the dilution of the sample is determined at which the odour is perceptible for 50% of the judges. This is in better accordance with the proposed definition of the Odour Number by Zoeteman and Piet (1974a). They define the ON as the dilution ratio of a sample at which the mixture of odorous components in water is present at its Odour Threshold Concentration (OTC). So:

$$ON = \frac{C_m}{(OTC)_m}$$

in which:

ON = Odour Number of the sample,
C_m = concentration of the mixture of odorants in the sample, expressed as fractions or multitudes of the OTC of the mixture,
$(OTC)_m$ = Odour Threshold Concentration in water of the mixture of odorants present in the sample.

According to this definition the ON value can also obtain values between 0-1 in which case less than 50% of the judges observe the odour. The way the Standard Methods define the Odour Number, its value can never become less than 1, which is not in accordance with reality. All samples containing absolutely no odorants up till those containing odorants at the OTC level are given an ON = 1, which is obviously incorrect.

It should furthermore be realized that in a series of ON measurements the ON-values obtained do not necessarily reflect the differences in perceived odour intensity. This aspect is often neglected in practice but can easily be deduced from the law of Stevens (1957) which describes odour sensation (S) as a power function of odorant intensity (I), or:

$$S = k \cdot I^n = k \left[\frac{C_m}{(OTC)_m} \right]^n = k \cdot (ON)^n$$

in which k and n are numerical constants, depending on the type of odorants concerned.

These equations show that ON-values can only be directly translated into odour sensation or odour intensity in case the samples concerned contain the same odorants or mixture of odorants. Otherwise samples with a same ON-value can in practice show large differences in perceived odour intensity, due to differences in the values of k and n. Therefore a direct method of scaling or cross-modality matching using undiluted samples are to be preferred for samples which change in chemical composition and which need to be compared over extended periods of time.

Finally it cannot be emphasized enough how important odour-free glassware and a good quality of the dilution water are for the reliability of the results. Particularly in the case of determinations of relatively low ON-values a poor quality of the dilution water can make the effect of all other precautions such as large panels, reduced influence of guessing, etc., negligible. This can be illustrated by the following example.

Suppose a water sample which actually should obtain an ON-value of 10 and a dilution water which contains the same odorant mixture as the sample at a fraction x of the OTC value. In this case the measured ON-value can be calculated for different values of x by the following equations:

$$A \cdot ON_A + B \cdot ON_B = (A + B) ON_{OTC}$$

and

$$ON_{measured} = \frac{A + B}{B}$$

in which:

A = volume of sample,
ON_A = theoretical ON value of sample,
B = volume of dilution water,
ON_B = ON value of dilution water,
ON_{OTC} = ON value of the sample diluted till the Odour Threshold.

In the case considered the following values apply:

ON_A = 10 . OTC
ON_B = x . OTC $(0 \leqslant x \leqslant 1)$
ON_{OTC} = OTC

and it can be derived that:

$$ON_{measured} = 1 + \frac{9}{1-x}$$

With this formula the following measured ON-values are obtained for different values of x:

$$x = 0.0 \quad ON_m = 10$$
$$x = 0.1 \quad ON_m = 11.0$$
$$x = 0.3 \quad ON_m = 14.0$$
$$x = 0.6 \quad ON_m = 24$$
$$x = 0.9 \quad ON_m = 92$$

This illustration shows that in case the dilution water contains similar odorants as the sample at 50% of the OTC the measured ON-value will be two times too high.

However, in case of the forced choice method the problem of dilution water, which is not completely odour-free, is less disturbing than in the case of the method of limits, because all the presented stimuli will have that odour and the difference between the odour of the sample and the odour of the blank is determined. A detailed description of the determination of the ON by the forced choice method is given by Grunt et al. (1977) and can be summarized as follows.

7.3.2.2 Determination of the Odour Number by the forced choice method

A series of dilutions of the water sample is made in freshly rinsed 100-ml erlenmeyer flasks, to be filled with 50 ml solution and covered by a glass cover. Two persons estimate the dilution at which the odour is just perceptible. Now a series of 7 dilutions is made, each differing by a factor of 3 in concentration and 3 dilutions being of higher resp. lower concentration than the estimated just perceptible dilution. The dilutions are made in duplo and all 14 stimuli are coupled with blanks. Within each of the 14 pairs (also a combination of 2 blanks and 1 dilution can be presented to further reduce the effect of guessing) the flasks are arbitrarily coded "M" and "V". The pairs are numbered from 1 to 14. The pairs are presented to about 15 panel members in randomized concentration at a temperature of 20°C. Each panel member smells the 14 pairs and judges for each pair which of the two has the strongest odour. A choice has to be made also when no perceptible differences in odour can be observed. For each pair the code (V or M) of the odour containing sample is noted. Judges may be situated in a circle and the pairs can be circulated with time intervals of 45 seconds. The percentage of correct responses is calculated and corrected for the 50% chance that the right flask has been indicated by the following formula:

$$R_{corrected} = \frac{R_{measured} - R_{chance}}{100 - R_{chance}} \times 100\%$$

in which:

$R_{corrected}$ = % right responses after correction for the chance of right guessing,
$R_{measured}$ = % correct responses measured,
R_{chance} = % correct responses due to guessing, which is 50% for the paired presentation considered.

Subsequently the ^{3}logarithm of the concentration (dilution) is plotted against the % of the corrected right responses and the value of the Odour Number is obtained from the curve by estimating the dilution value which corresponds with a value of $R_{corrected} = 50\%$. A more accurate estimation of the ON-value can be obtained by converting the $R_{corrected}$ values to z-scores.

Generally it was found for ON determinations of river water samples (Grunt *et al.*, 1977) that by using panels of 15 persons, this method results in an accuracy of the **Odour Number of 25% (90% probability of exceedence). The calculated accuracy is** 35% for a panel of 10 persons and 55% for a panel of 5 persons. Therefore it is felt that the described Standard Methods procedure, which is based on a minimum number of only 5 judges and the method of limits, gives rather unreliable results for most purposes.

7.3.2.3 *Other methods for routine odour intensity estimation*

The described ON determination by the forced choice method is relatively time consuming. Other more rapid methods have therefore been suggested for routine applications. Water chemists of course first try to identify those compounds responsible for water odour and try to develop routine analytical methods for the most relevant odorants. This may be successful in certain cases where the chemical composition of the water is relatively constant and where the odorants can be easily identified and quantified. An example in this respect is discussed later in relation to taste assessment of drinking water. Generally this approach cannot be followed. Under circumstances where samples are more or less alike but suitable chemical analysis is not available the odour intensity of undiluted samples can be assessed on an interval scale. As stated in the preceding paragraph, this direct scaling method has the advantage of producing results which are better correlated with the actual perceived odour strength. On the other hand, the statistical reliability of the direct scaling method is inferior to the ON measurement described previously and it is only usable for a series of samples within a certain intensity range and with a similar odour character.

Good experiences were obtained for surface water samples (Grunt *et al.*, 1977) using a 10 point interval scale from "not perceptible" (0) up to "very strong" (10) and standard reference solutions of isoborneol in water containing, 5, 50 and 500 µg/l respectively, of isoborneol. These isoborneol solutions were defined as corresponding with, the values 2, 5 and 9 respectively, on the interval scale.

Before presentation the samples are diluted if necessary to obtain an odour strength which is comparable to those of the isoborneol reference solutions. All solutions are offered in 100-ml erlenmeyer flasks provided with a glass cover and filled with 50 ml of the stimulus. Panel members were asked to assign a scale value after comparison with the isoborneol reference solutions. A series of undiluted samples had to be presented 4 times to a panel of 15 judges to obtain results with a similar accuracy as the ON determination described in paragraph 7.3.2.2.

Based on the results obtained with both methods for 16 river water samples a linear relationship between the scale value and log(ON) was found as shown in figure 7.1. In this way it became possible to estimate ON values for surface water samples in 20-30% of the time needed for ON measurements by the forced choice method. However, in case samples were containing odorants of a different odour character large deviations are introduced.

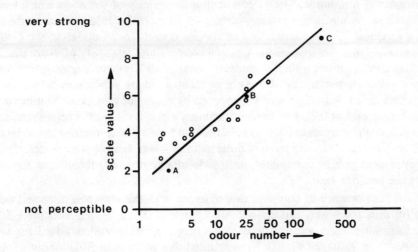

Figure 7.1
Relation between Odour Numbers and scaled odour intensity for Dutch river samples and three isoborneol-solutions. (A,B,C, resp., containing 5,50 and 500 µg/l isoborneol)

It was found, for instance, that very unpleasant odours were mostly scaled too high because intensity and preference were not distinguished sufficiently. It is recommended for each series of similar types of water to establish a curve as shown in figure 7.1 and to check regularly its validity.

7.3.3 Determination of taste of drinking water

7.3.3.1 *General aspects of the taste determination and standards for drinking water taste*

Before discussing the possible methods of taste measurements in more detail it is necessary to pay attention to the existing and proposed standards for water taste. In Chapter 5 a number of reasons are presented why individual chemicals should not be present in drinking water in concentrations above 1% of the OTC. One of the reasons was that the most sensitive 5% of the population may still be able to detect those levels of contamination.

Furthermore, it is necessary to develop a criterion for the maximum acceptable level of taste impairment relating to the overall drinking water quality. The WHO International Standards for Drinking Water (1971) define the maximum permissible level of substances causing taste and odour as the level at which their presence is *"unobjectionable"*.

In a report of the Committee on Water Quality Criteria of the National Academy of Sciences in the USA (US EPA, 1973) it is stated that *"public water supply sources should be essentially free from objectionable odour"*. In a directive of the Council of the European Communities (1975) concerning the quality of surface water intended for the production of drinking water, certain odour intensity guide values are described. In a proposal of the Commission of the European Communities to the Council (1975), relating to a directive of the Council concerning the quality of water destined for human consumption, a similar quantitative approach is given to perceptible water quality aspects. A maximum allowable concentration of substances which impair taste and odour of drinking water is proposed by means of an Odour Number and a Taste Number of 3 at 25°. For both parameters a guide value of zero is mentioned. Although not explicitly stated it seems that the odour and taste number has to be considered as the required number of dilutions of a water sample with water, free of odour or taste impairing substances, until 50% of a number of subjects can no longer observe the smell or taste.

It is of importance to compare this criterion with the obtained taste and odour ratings for drinking water, as described in Chapter 2. As can be seen from table 2.7 all types of drinking water will easily meet the proposed European standard for water odour. Also for water taste it can be concluded that more than 50% (about 65%) of the observers rated the taste of drinking water derived from stored surface water as *"good"* or *"not perceptible"*. So it cannot be expected that this type of proposed standards for drinking water in the European Communities will contribute to an improved perceptible quality of drinking water. From the quoted references it is clear that water taste should not be objectionable. However, the percentage of consumers to which this requirement should apply is not specified. In view of the results of this study it is recommended that only less than 5% of the population served by a public water supply system, may assess water taste as objectionable (unpleasant/offensive) or worse. This would mean, as table 2.7 indicates, that all ground water supplies studied by means of the national inquiry would easily meet the recommended standard but that an improvement of water taste is needed for those supplies using surface water as a raw water source.

The results discussed in paragraph 7.3.2 indicate that the normal determination of a Taste Number, similar to the ON determination, is less suitable to detect the unwanted presence of odorants at sub-threshold levels. For this purpose hedonic taste rating by large laboratory panels can be considered. However, as stated by Bruvold (1970), "values representing . . . quality ratings for a laboratory panel are only estimates of these values for the entire consuming population". Therefore the possibilities of using large consumer panels need to be explored further. Finally, attention will be paid to the possibilities of direct chemical determination of water constituents which are indicators for impaired waste water.

7.3.3.2 *Taste assessment by laboratory panels or by consumer groups*

As discussed in Chapter 3 taste assessment based on a taste rating test by a laboratory panel or by means of an inquiry among consumers may result in large differences in outcome as consumers tend to be less critical than panel members. Generally a suitable method for systematic determinations of drinking water taste has to meet the following requirements:
1. A representative number of different tapwater samples of a water supply system has to be considered to include cases of impaired water taste due to distribution.
2. The time between tapwater sampling and sensory water quality assessment should be as short as possible.
3. A representative number of individuals must be included with respect to factors such as age, distribution of odour sensitivity, etc., to obtain data applicable to the population supplied.
4. The persons selected should have basic experience with the test procedures.
5. A sufficient number of observations for one water sample must be collected to allow discrimination between accidental differences and those due to changes in the quality of the finished water.
6. Total cost should be sufficiently low to permit water works to use the method at least on a weekly basis.

Based on these requirements it is suggested to use a group of consumers instead of a laboratory panel for the regular sensory assessment of drinking water quality.

Consumer panels can be selected in a similar way as described in paragraph 3.2.1. Participants should be instructed and motivated before and during the testing. It is recommended to use a panel of 100 consumers for each area supplied by a water supply system. Panel members should weekly assess water taste at their homes and report the result on a category scale as suggested in the Standard Methods or in table 3.6. In this way reliable estimates of water taste assessment by the population supplied can be obtained and changes in water quality which need additional correction at the treatment plant can be registered.

Although similar data can be obtained with large laboratory panels, this procedure will be much more time consuming. It can be estimated that to obtain a similar type of accuracy by means of laboratory panels as by the panel of 100 consumers samples should be presented, for example, 10 times to a panel of 10 judges. The estimated accuracy of the measured taste scale value according to the scale presented in table 3.6 is in this case 7% ($\alpha < 0.05$) (see also table 3.9). In case samples are presented 10 times or 3 times to a panel of 5 judges the estimated accuracy becomes 18% and 33% respectively. To be able to distinguish between water quality changes that need correction an accuracy of about 5% should be strived after. Confounding factors can, of course, be better controlled in laboratory panels than in consumer panels, and may finally be decisive for the choice.

Of course detailed investigations which are incidentally needed to study treatment processes or new water sources can best be performed with laboratory panels. In that case it is further recommended to apply the forced choice method. The dilution water should preferably not be produced artificially but obtained from a natural water source in the neighbourhood.

7.3.3.3 *Estimation of water taste intensity by measuring chemicals*

Using the data described in table 3.9 and appendix 5.1, it can be investigated if individual chemicals or groups of chemicals in drinking water can be found which can be considered as general indicators for water taste. Correlations were calculated between average water taste ratings and the presence of 7 groups of organic compounds, compiled from appendix 5.1. The results are given in table 7.1.

Table 7.1: Correlation between average water taste rating and the presence of groups of organic micropollutants in tapwater

Groups of organic compounds	Correlation coefficient (r)
Total haloforms	0.45
Total alkylbenzenes	0.47
Total polycyclic aromates	0.17
Total phtalates	0.27
Total chlorobenzenes	0.56
Total chlorinated ethers	0.63
Total anilines	0.55

The data of table 7.1 suggest that water taste is associated with all parameters given except for the polycyclic aromates and phtalates. A more specific multiple regression analysis was made for taste rating and the presence of representative compounds for the groups of haloforms, chlorobenzenes and chlorinated ethers. This choice includes the influence of anilines and alkylbenzenes as these two groups are highly correlated with the group of chlorinated ethers (r is 0.88 and 0.89 respectively,). As chloroform (r = 0.45), o/p-dichlorobenzene (r = 0.61) and bis (2-chloroisopropyl)ether(r = 0.59) each showed high correlation coefficients with taste rating, these were finally chosen as representative compounds, resulting in the following regression equation:

$$\text{Taste rating} = 1.26 + 0.01 X_1 + 1.83 X_2 + 0.26 X_3$$

in which:

X_1 = concentration of chloroform (μg/l),
X_2 = concentration of o- and p-dichlorobenzene (μg/l),
X_3 = concentration of bis(2-chloroisopropyl) ether (μg/l).

This equation accounts for 78% of the variance of the taste ratings. To increase the predictive value one extreme point out of the 20 was excluded and the following non-linear model was adopted, which is better in accordance with the fact that sensory sensation grows as a non-linear function of the odorant concentration:

$$\text{Taste rating} = {}^e\log(0.053X_1 + 1.41X_2 + 7.2X_3 + 3.4)$$

With this equation the taste rating of the 19 types of water considered can be predicted with an accuracy of 11% as further illustrated in figure 7.2.

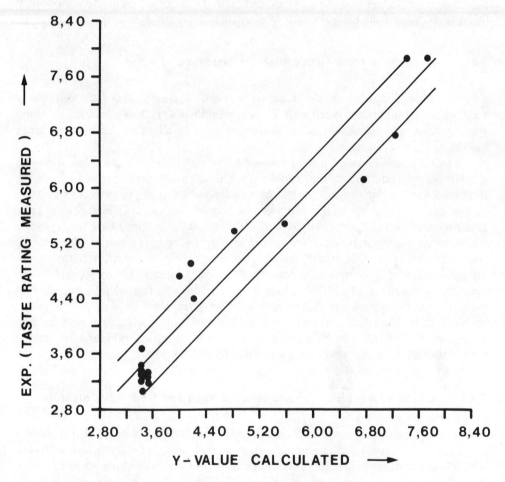

Figure 7.2.
Multiple linear regression of exp. (taste rating) on chloroform, dichlorobenzenes and bis(2-chloroisopropyl)ether for 19 types of drinking water in The Netherlands.

This result strongly supports the view that taste of drinking water can be measured in a reliable and objective way and that this water quality determination can be used as a valuable overall indicator for water quality. The surprising results further suggest that in the case of The Netherlands drinking water taste could reliably be predicted by measuring routinely chloroform, dichlorobenzenes and bis(2-chloroisopropyl)-ether which analysis can be carried out simply in one GC-run. It must, however, be realized that such a simple relation will probably be only valid for a certain period and in a limited geographic area. Therefore it is expected that only in exceptional cases measurement of a few chemicals will be representative for the occurrence of taste problems. In view of the more universal value of taste rating as an indicator parameter it is recommended to use consumer panels for water quality evaluation over longer periods which determination can occasionally and temporarily be replaced by routine measurement of a small number of chemicals.

7.4 Measures to reduce perceptible contamination of water

The data described in the Chapters 2 and 3 clearly illustrate that many waterworks are nowadays faced with the nearly impossible task of preparing drinking water, which is tasteless and odourless for practically all of the consumers, from heavily contaminated sources.

As shown in Chapter 4 only a few inorganic constituents of drinking water do actually cause taste and odour problems. Among those cations which are most important in taste impairment, magnesium particularly may slightly affect water taste in the case of ground water supplies. Generally the large differences in taste assessment of water distributed by surface water supplies compared with ground water supplies cannot be attributed to inorganic water constituents. Measures to overcome problems of impaired water taste should primarily aim at reducing the introduction of the compounds concerned at the source. The same applies for problems of tainting of fish or odour problems directly related to contaminated surface water or waste water. Therefore the cause of impaired smell or taste of water should first be identified and tackled. For those substances which can only at high cost or in the long term be eliminated at the source additional measures to remove these compounds at the treatment plant have to be applied.

7.4.1 Causes of impaired taste and smell of water and fresh water organisms

A main cause of taste and odour problems is the presence of perceptible chemicals in industrial and municipal waste waters discharged into surface waters in which fish and other aquatic organisms live and from which drinking water has to be derived. In the Freshwater Aquatic Life and Wildlife Section of the report on Water Quality Criteria (U.S. EPA, 1973) a survey is given of concentrations of chemical compounds in water which can cause tainting of the flesh of fish and other aquatic organisms. A summary of the most odour intensive compounds is given in table 7.2. Comparison of the data of table 7.2 with OTC data reported by Gemert and Nettenbreijer (1977) suggest that the estimated threshold levels given in table 7.2 are probably too high.

Table 7.2: Concentrations of some chemical compounds in river water that can cause tainting of the flesh of fish (from U.S. EPA, 1973)

Compound	Estimated threshold level in water ($\mu g/l$)
Acetophenon	500
n-Butylmercaptan	60
4-tert. Butylphenol	30
2-Chlorophenol	1
4-Chlorophenol	10
4-Cresol	100
1,2-Dichlorobenzene	200
2,4-Dichlorophenol	3
2,6-Dichlorophenol	30
Diphenyl oxide	50
Ethanethiol	200
Ethylacrylate	600
Guaiacol	80
2-Methyl, 6-chlorophenol	3
Naphthalene	1000
2-Naphthol	300
α-Methylstyrene	250
4-Quinone	500
Styrene	250
2,4,6-Trichlorophenol	10

A similar list of organic compounds of industrial nature which can cause taste and odour problems in drinking water is presented in table 7.3.

Table 7.3. Concentrations of some chemical compounds in river water that can cause impaired taste of drinking water

Compound	OTC ($\mu g/l$)	Maximum concentration detected in Rhine water at Lobith in 1978 ($\mu g/l$)	Recommended maximum allowable concentration in surface water ($\mu g/l$)
Biphenyl	0.5	0.1	0.01
Bis(2-chloroisopropyl)ether	300	1.3	10
2-Chloroaniline	3	0.3	0.1
2-Chlorophenol	0.2	2.3	0.01
4-Chlorophenol	0.5	3.9	0.01
5-Chloro-o-toluidine	5	0.5	0.1
3,4-Dichloroaniline	3	0.1	0.1
1,2-Dichlorobenzene	10	3.2	0.1
1,4-Dichlorobenzene	0.3	3.6	0.01
2,4-Dichlorophenol	2	0.6	0.1
2,6-Dichlorophenol	3	0.5	0.1
Hexachlorobutadiene	6	2.4	0.1
β-Hexachlorocyclohexane	0.3	0.1	0.01
Naphthalene	5	1.8	0.1
1,2,4-Trichlorobenzene	5	2.2	0.1
1,3,5-Trimethylbenzene	3	0.2	0.1

The data listed in table 7.3 are based on the results presented in Chapter 5 as well as recent data from Wegman *et al.* (1978) and Morra *et al.* (1979). The philosophy behind the recommended maximum concentrations in surface water for these compounds is given in the next paragraph.

Many other compounds can be involved in local odour nuisance problems due to industrial discharges of waste water into surface water. Examples are organic sulphur compounds formed during anaerobic transport of waste water, such as 4-methyl-4 mercaptopentan-2-on from mesityloxide and constituents of oil wastes like indenes, naphthalenes, terpenes, etc.

A second category of causes of the introduction of odorants is the biological formation during exposure of eutrophic water to light or as a result of growth of microorganisms in the distribution system. Within this category the early smelling geosmin

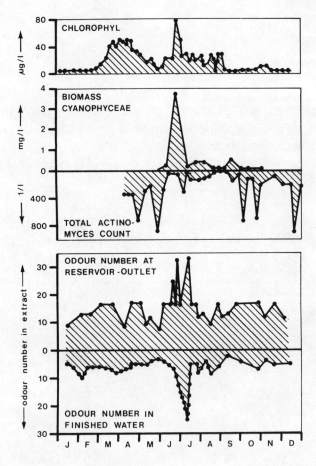

Figure 7.3
Odour development by biological growth in the reservoir "Grote Rug" of the waterworks of Dordrecht, The Netherlands, in 1973.

and 2-methyl-isoborneol are important as well as a number of aldehydes such as decanal. These compounds can be formed during the stay of water in lakes or reservoirs or during transport of finished water at temperatures above 15°C. A typical example of odour problems as a result of growth of blue-green algae in a storage reservoir of the waterworks of the city of Dordrecht is given in figure 7.3 (Zoeteman and Piet, 1974b). The ON-values were in this case measured in concentrates of the reservoir water, which is the reason for the high levels indicated. The real ON-values were roughly a factor 2 lower.

Finally water treatment itself can be the cause of the presence of certain odorants in drinking water. Besides the well-known example of the formation of chlorophenols during chlorination of phenol containing waters as well as the production of haloforms from humic and fulvic acids (Rook, 1974), it should be realized that in general chlorinated substitutes of aromatic and aliphatic substances have OTC values which are considerably lower (100-1000 times) than the original substances. Therefore removal of organics will result in a better water quality than changing the compounds by chemical or biological processes. Also the use of coatings of metal surfaces or plastic pipes in water supply systems have led to release of organic substances from these materials into the water and subsequent taste problems.

7.4.2 Prevention and reduction of taste problems within water supply systems

Of course the design of the water supply system should be such that taste and odour problems are avoided as much as possible. In this respect the following factors are important:

—use the least polluted raw water sources available, such as ground waters instead of surface waters;
—avoid exposure of the water to light, particularly in case of waters containing high levels of phosphorous and nitrogen compounds to limit growth of algae, by storing water in the underground instead of open reservoirs;
—avoid the use of chlorine;
—avoid high temperatures, high levels of residual organic carbon content and long residence times in the distribution system in order to prevent biological growth during water distribution;
—use coatings and piping materials which do not release perceptible compounds.

For the removal of odorous chemicals from contaminated raw waters ozone has been used in different cases successfully in Europe. Generally, however, ozone, applied at a dose of 1-3 mg O_3/l, will reduce the TN or ON-value only by a factor of 2-4, as shown in figure 7.4 for 7 ozone installations in The Netherlands (Meijers, 1977). Substances like the higher chlorinated benzenes, nitrobenzenes as well as lindane, hexachlorobutadiene and tetrachloroethene are hardly oxidized by ozone. Anilines and chlorophenols are more easily affected (Kruithof, 1978).

Application of ozone in combination with a conventional water treatment system will therefore not always produce tasteless drinking water. A further illustration is given in table 7.4 for the case of the water supply of Dordrecht in 1973 (Zoeteman and Piet, 1974a).

Table 7.4. Average ON-values in 1973 during treatment of river Rhine water at Dordrecht

Stage of treatment	ON
Rhine water at reservoir inlet	30
Reservoir (T = 100 days) outlet	15
After breakpoint chlorination and coagulation	12
After ozonation	3

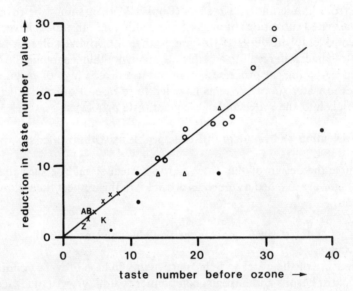

Figure 7.4.
Reduction in Taste Number value by ozonation as a function of the Taste Number before ozonation for 7 installations in The Netherlands (after Meijers, 1977)

Besides ozone, activated carbon adsorption is increasingly applied to reduce taste and odour problems and to remove organic chemicals in general. Up till the sixties dosing of powdered activated carbon was commonly used for taste and odour control as illustrated in paragraph 1.3.1. High doses up till 50 mg/l were often applied without success (Hansen, 1972). The use of granual activated carbon filters allowed for a much better result (McCreary and Snoeyink, 1977). In the case of moderately contaminated waters fresh carbon filters will remove all taste and odour compounds during a period of 3-6 months. In practice filters are often used over a period of 1-2 years before the carbon is replaced. Contaminated water with ON-values of 100 or more and odorant levels above 10 μg/l will generally result in much shorter periods of 1-2 months after which breakthrough takes place. Chlorinated benzenes are strongly adsorbed on all types of carbon and are very little displaced. Relatively volatile chlorinated alkanes are best removed by hydrophobic carbons and belong to the group of compounds breaking through first. More polar substances like bis(2-chloroisopropyl)ether show an intermediate behaviour. The better results with granular carbon filters are not only due to the better use of the adsorptive capacity of the carbon, but also to biological degradation which occurs in the upper part of the filter.

The combined use of ozone and carbon filters will provide in most cases a more or less acceptable water taste when contaminated surface waters have to be used. Ozonation before carbon filtration will result in increased effluent bacterial counts which necessitates frequent backwashing and post-disinfection. According to Hrubec (1978) the Taste Number of Rhine water after coagulation and filtration improved by ozonation from 12 to 4.5 in 1975 and subsequent carbon filtration resulted in an average TN-value of 2. Similar data are reported by Kühn *et al.* (1978) showing a reduction in Dissolved Organic Chlorine content of bankfiltered Rhine water from 61 to 45 $\mu g/l$ due to ozonation and subsequently to 27 $\mu g/l$ after activated carbon filtration.

Recently also ion-exchange resins have been studied for removal of organic compounds (Anderson and Maier, 1979).

Summarizing, it must be concluded that advanced treatment techniques can often reduce but not completely eliminate taste and odour problems.

7.4.3 Abatement of the presence of taste and odour impairing substances in surface waters

Abatement of the problem of tainting of fresh water fish and of impaired drinking water taste, due to organic chemicals which cannot be sufficiently removed even by advanced water treatment systems, has to be realized by better controlling the discharge of compounds such as those listed in tables 7.2 and 7.3, and by controlling the nutrients supporting growth of blue-green algae and streptomycetes.

The taste and odour impairing substances are part of the so-called "grey list" of compounds of a directive of the Council of the European Communities (1976) and of the Convention against the chemical pollution of the river Rhine (Koninkrijk der Nederlanden, 1977). This implies that these compounds are given less priority in sanitation than the compounds of the "black list". The findings presented in this study, however, emphasize the need to reduce the discharge of these chemicals in the near future to enable a proper functioning of public water supply systems and a better quality of fresh water fish. Water quality objectives for these types of water use should be based on a similar advanced rational as presented for drinking water taste in paragraphs 5.4.1 and 7.3.3.1. The criterion suggested was that not more than 5% of the consumers may detect the taste of a compound in drinking water. A similar approach was recommended by Bruvold (1977) who emphasized that standards for odorants should be below the OTC-value, also from the point of view that subthreshold mixtures of taste and odour producing chemicals are likely additive and possibly even multiplicative. This criterion of 5% corresponds with a highest acceptable concentration of the compound in drinking water of 1% of the OTC-value (paragraph 5.4.1). As odorants may easily pass water treatment systems and raw water quality should be such that drinking water can be produced without the need of applying advanced techniques like ozonation and carbon filtration, it is suggested that levels of odorants in surface waters are not allowed to be more than a factor 3 higher than in drinking water. Based on this general principle, that taste and odour causing chemicals should not be present in surface waters destined for the production of drinking water at levels above 3% of their OTC, the recommended water quality objectives presented in table 7.3 were derived. It is up to the river water authorities to translate such objectives to emission standards. Assuming a general dilution factor of 100 it is suggested that relatively persistent odorous chemicals should not be present in effluents at levels exceeding their OTC 3-10 times.

Part of the measures needed to reduce odours due to photosynthesis in stagnant water are similar to those envisaged in eutrofication control, including replacement of phosphates in detergents by chemicals like NTA and phosphorous removal at the municipal waste water treatment plants. Algal control in open storage reservoirs can be further obtained by promoting mixing in reservoirs with a depth of 10 or more metres or by removing phosphates at the reservoir inlet by coagulation and sedimentation.

7.5   Final considerations

It is recommended to continue the identification of organic compounds in surface water, fish and drinking water and to collect additional information in the area of OTC and TTC data for compounds already identified. Furthermore, it is necessary to investigate the combined olfactory and gustatory effects of inorganic and organic compounds in order to evaluate the contribution of individual compounds to the final taste of water and fish.

As stated by Bruvold (1977): *"Enough progress has been made in consumer and in threshold research to encourage further work on both, toward attaining the best possible perceptual water quality for all consumers"*. A prerequisite for such activities is, however, the willingness to mobilize much larger groups of judges in laboratory panels than the present habit of only using 2-5 "expert noses" and to carry out regular inquiries among large groups of consumers. The usefulness of such an advanced approach can no longer be ignored. The cost involved is rather low compared with the high cost of modern laboratory instrumentation. The results which are obtained have a direct meaning for the water user. Sensory assessment can provide a cheap warning mechanism for many cases of accidental pollution. There is a growing awareness of the potential health risks related to water pollution. Substances are involved which may accumulate in the waters due to direct and indirect reuse practices in water short areas such as the United Kingdom, Japan, South Africa, Israel, California and other densely populated areas in the world (WHO-IRC, 1975) (English et al., 1977) (Zoeteman, 1977) (Shuval, 1977). It may be expected that, due to the growing water demand, these reuse practices will gradually spread over the entire world. More and more refractory industrial compounds have been detected at low concentrations in drinking water which are mutagenic, teratogenic or carcinogenic. The public starts to develop a critical attitude towards many achievements of modern society. More and more persons rediscover the merits of long known natural processes and principles.

Sensory water quality assessment is the only means by which the consumer can evaluate water quality personally and instantaneously. In the past a bad perceptual water quality has often resulted in effective counter-measures from a health point of view even when the cause of the problems was not well understood. The Metropolis Water Act of 1852, which prohibited direct abstraction of river Thames water and which prescribed the use of slow sand filtration in the district of London around St. Paul's Cathedral, was based on the objective of obtaining water of better perceptual quality. The Act resulted in a much lower cholera frequency at a moment that the science of bacteriology was not yet founded.

The differences in taste and odour of the types of drinking water described in this work point at shortcomings in our present water supply systems which need correction. Examples are a better control of discharges of odorous chemicals, avoidance of circumstances which permit biological growth, and more advanced treatment of contaminated raw water sources such as river bank filtrates which are sometimes unjustly considered as good quality ground waters.

Against this background it may be expected that in the near future gustation and olfaction will be more seriously applied than in the past decades, both by consumers and water authorities.

Where water is scarce, even a few drops become precious

Summary

In the past, sensory assessment was the most important means by which the quality of drinking water was evaluated. Recently increasing numbers of contaminants of drinking water can be measured by instrumental techniques. For this reason the sensory assessment is given only little attention in water quality analysis. The consumer of drinking water, however, is still evaluating water quality mainly by his senses.

The main purpose of this study is to investigate the suitability of the sensory assessment of water quality as an indicator for the presence of chemical contaminants in drinking water and for potential health effects on consumers. Among the sensory quality aspects, which include transparency, colour, temperature, odour and taste, especial attention is given to the taste and smell.

This investigation attempts to answer the following questions in particular:
- which general causes and which individual compounds can be responsible for bad taste and odour of drinking water;
- to which extent do the senses of taste and smell function as a warning system for possibly hazardous levels of chemical contaminants in drinking water;
- how can taste and odour of drinking water be measured in a reliable way and on a routine basis by water works;
- which standards for sensory water quality aspects should be applied in view of a function as an overall indicator for contamination and based on health considerations?

Chapter 1 gives a description of the aim of this investigation, as well as an introduction relating to the nature of sensory assessment of water quality. Furthermore the literature on perceptible substances detected in drinking water is summarized.

Chapter 2 presents the results of an inquiry which was held among a sample of the population of The Netherlands consisting of 3073 persons of 18 years and over. Ratings for the quality of the drinking water in the category "offensive" or a worse category were given by 1.3% of the persons for colour and turbidity, by 3.2% for odour and by 6.9% for taste of their drinking water. Water temperature was rated as "tepid" or worse by 0.2% of the individuals. The quality of drinking water derived from ground water was significantly better than that of water derived from surface water sources, in relation to taste, odour and tea flavour. Increased levels of magnesium of about 30 mg/l seem to affect the taste of water derived from ground water. Water taste could be shown to be the major factor determining the assessment of the acceptability of drinking water quality.

In order to identify the compounds causing bad taste and odour of drinking water, 20 types of drinking water were collected. In Chapter 3 the sensory assessment of the quality of these tapwaters by a selected panel of 52 judges is discussed. Generally, the results confirmed those obtained by the national inquiry. As the taste of drinking water is much better noticeable than the smell, assessment of the quality of drinking water by the chemical senses can be limited to tasting. The taste of the water distributed by the 8 ground water supplies studied was considerably better, as indicated by a 5-category taste scale, than the taste of 5 types of drinking water prepared from surface water, which was infiltrated in the dunes. The highest average taste scale value was found in the case of the 7 supplies using surface water after storage in open reservoirs or after bank filtration. Additionally the experiments with the panel resulted in a number of recommendations to be considered in the selection and application of panels for taste assessment of water. It was also found that panel members generally rated the taste of the tapwater at their homes much more positive than the taste of comparable types of water presented during the panel sessions.

The possible contribution of inorganic water constituents to impaired taste is considered in Chapter 4. The present levels of sodium, calcium and magnesium salts in water derived from surface water could not explain the large differences in taste between these water types and types of water derived from ground water. Therefore certain organic substances must be the cause of the observed differences in taste. Comparable levels of *"offensive"* taste impairment are tentatively estimated at 100 mg/l for the calcium-ion and 175 mg/l for the sodium-ion on basis of dilutions of chloride, hydrocarbonate and sulphate salts of these cations in distilled water. It is generally concluded that inorganic constituents of drinking water at the levels detected contribute more to the desirable neutral taste of drinking water than to taste impairment.

As described in Chapter 5, the analysis of organic compounds in the 20 types of drinking water studied resulted in the detection of 280 organic substances. Most of the compounds were detected by a gas chromatograph-mass spectrometer-computer system. The results obtained represent only part of the mixture of compounds present in drinking water. Nearly twice as much organic compounds were found in the case of surface water supplies in comparison with ground water supplies. The highest number of organic compounds was found in tapwater derived from Rhine water after storage in open reservoirs. This was probably due to the combined effect of the formation of compounds by organisms and by break-point chlorination.

For drinking water derived from surface water a number of taste impairing compounds of possibly biological nature were selected, including the earthy smelling geosmin and 2-methylisoborneol as well as octene-1, 2-ethylhexanol, nonanal and decanal. Furthermore a number of taste impairing compounds of industrial origin could be detected in relation to surface water supplies, among which are dichloro- and trichlorobenzenes, chloro- and dichloroanilines, 5-chloro-o-toluidine, hexachlorobutadiene and bis(2-chloroisopropyl) ether. The compounds are probably the cause of the chlorine-like taste quality of certain waters derived from bankfiltrate of Rhine water. Drinking water derived from ground water often contained certain esters like methyl isobutyrate and methyl 2-methylbutyrate which might contribute to a pleasant taste of those waters. However also certain chlorinated chemicals, like tetrachloromethane and trichloroethene are present in ground water supplies, sometimes well above the microgram/litre level.

Sensory assessment of drinking water quality and health protection aspects are discussed in Chapter 6. Based on water consumption data, obtained from the inquiry mentioned earlier, it can be shown that water taste assessment is an important factor in the quantity of drinking water consumed as such. However in the case of tea, coffee and soup, even an offensive water taste did not result in a reduced consumption.

The total quantity of tapwater consumed amounts to 1.36 litre/head/day for those adults assessing the taste as good and to 1.17 litre/head/day in the case of an offensive or worse taste. It is concluded that chemoreception of water contaminants functions only in a rudimentary and limited way as a mechanism in health protection. Furthermore a literature survey showed that the odour detection thresholds for hazardous chemicals, exerting long term toxic effects, are not sufficiently low and specific to prevent unnoticed exposure to levels exceeding the maximum acceptable intake for certain compounds such as aldrin, dieldrin and lindane. On the other hand no chemicals are presently known for which consumption by an adult person within 24 hours of 2 litres of water, containing the chemical in a concentration below the Odour Threshold Concentration, would probably result in acute lethal effects. Absence of bad taste or smell, does not guarantee the safety of the drinking water.

Finally, Chapter 7 describes a number of applications and recommendations. It is recommended to assess sensory water quality aspects at least weekly in the case of surface water supplies by carrying out an inquiry among the consumers situated in the area served. It is further recommended that less than 5% of the population served by a public water supply system may assess water taste as offensive, unpleasant or worse. Practical methods for taste and odour determination are described and commented on.

The data presented illustrate that many waterworks are facing a nearly impossible task to prepare drinking water, which is tasteless for practically all of the consumers, from heavily contaminated surface water. Similar problems arise in relation to fresh water fish. To improve this situation a sanitation program for the compounds concerned should at least result in a reduction of the levels of taste impairing substances in river water below 3% of the Odour Threshold Concentration and in the effluents below levels of 3-10 times the Odour Threshold Concentration. In the meantime waterworks should strive for a better removal of organic substances from the contaminated raw water by application of ozonation and particularly granular carbon filtration. They also should avoid as much as possible algal growth, as this results in biological production of organic taste impairing compounds. In this respect surface water storage by means of artificial recharge in the underground should be preferred to open storage reservoirs. For existing reservoirs a reduction of the level of nitrogen and phosphorous compounds (plant nutrients) in the raw water should be strived for. Except for the limitations mentioned in Chapter 6, water taste can be considered as a useful overall indicator for chemical contamination and insufficient treatment of drinking water. Therefore perceptible contamination of drinking water should be given more attention by those responsible for water quality control.

List of symbols, abbreviations and synonyms

Frequently used symbols and abbreviations:

α	One-tailed probability of exceedence of a coefficient of correlation
ADI	Acceptable daily intake (mg/kg/day)
C	Concentration of an organic water constituent (μg/l)
CBS	Centraal Bureau voor de Statistiek (Central Bureau for Statistics), Voorburg, The Netherlands
d	Day
ECD	Electron capture detector
EX	Liquid-liquid extraction method
FID	Flame ionization detector
g	Gram
GC	Gas chromatograph
h	Head
H	Measure of association between an ordinal and a nominal variable, Kruskal-Wallis test
HS	Head space
IUPAC	International Union of Pure and Applied Chemistry
l	Litre
LD_{50}	Lethal dose in 24 hours of a chemical for 50% of experimental animals (mg/kg)
m	Milli-
μ	Micro-
meq	Milli-equivalent
MS	Mass spectrometer
NIWS	National Institute for Water Supply (Rijksinstituut voor Drinkwatervoorziening, RID), Voorburg, The Netherlands
ON	Odour number
OTC	Odour threshold concentration in water (μg/l)
p	One-tailed probability of exceedence of a measure of association
r	Coefficient of correlation
t	Test statistic for the t-test
TLC	Thin-layer chromatographic method
TN	Taste number
TTC	Taste threshold concentration of an organic compound in water (μg/l) or an inorganic constituent of water (mg/l)
WHO	World Health Organization, Geneva, Switzerland
[X]	Concentration of a water quality parameter (mg/l)
z	Measure of association between two ordinal variables, standard normal deviate

Chemical synonyms of substances mentioned frequently in the text:

Acrylonitrile	Propenoic acid nitrile
Aldrin	1,2,3,4,10,10-Hexachloro-6,7-epoxy-1,4,4a,5,6,7,8,8a-octahydro-*exo-exo*-1,4:5,8-dimethanonaphthalene
Alkylmercaptan	Alkanethiol
3,4-Benzofluoranthene	Benzo[*b*]fluoranthene
11,12-Benzofluoranthene	Benzo[*h*]fluoranthene
1,12-Benzoperylene	Benzo[*pqr*]perylene
3,4-Benzopyrene	Benzo[*a*]pyrene
Bis (2-chloroisopropyl) ether	Bis (2-chloro-1-methylethyl) ether
Bromoform	Tribromomethane
sec-Butylamine	2-Aminobutane
Chlordane	1,2,4,5,6,7,8,8a-Octachloro-3a,4,7,7a-tetrahydro-4,7-methanoindane
Chloroform	Trichloromethane
5-Chloro-*o*-toluidine	2-Amino-5-chlorotoluene
Cinnamaldehyde	3-Phenylpropenal
2,4-D	2,4-Dichlorophenoxy acetic acid
p,p'-DDE	1,1-Dichloro-2,2-bis (*p*-chlorophenyl)ethene
p,p'-DDT	1,1,1-Trichloro-2,2-bis(*p*-chlorophenyl)ethane
Dieldrin	1,2,3,4,10,10-Hexachloro-6,7-epoxy-1,4,4a,5,6,7,8,8a-octahydro-*endo-exo*-1,4:5,8-dimethanonaphthalene
1,1-Dimethoxyisobutane	1,1-Dimethoxy-2-methylpropane
Endrin	1,2,3,4,10,10-Hexachloro-6,7-epoxy-1,4,4a,5,6,7,8,8a-octahydro-*exo-exo*-1,4:5,8-dimethanonahpthalene
Geosmin	*trans*-1,10-Dimethyl-*trans*-9-decalol
Haloforms	Trihalomethanes
γ-HCH	γ-1,2,3,4,5,6-Hexachlorocyclohexane
Heptachlor	1,4,5,6,7,8,8-Heptachloro-3a,4,7,7a-tetrahydro-4,7-methanoindene
β-Hexachlorocyclohexane	β-1,2,3,4,5,6,-Hexachlorocyclohexane
Lindane	see γ-HCH
Malathion	O,O-dimethyl-S-[1,2(di(ethoxycarbonyl))ethyl]dithiophosphate
Methoxychlor	2,2-Bis (4-methoxyphenyl)1,1,1-trichloro-ethane
2-Methylisoborneol	2-*exo*-Hydroxy-2-methylbornane
Methyl isobutyrate	Methyl 2-methylpropanoate
Methyl 2-methylbutyrate	Methyl 2-methylbutanoate
Nicotine	1-Methyl-2-(3'-pyridyl)pyrolidine
Parathion	O,O-diethyl-O-(*p*-nitrophenyl)thiophosphate
2,3-Phenylenepyrene	Indeno[1,2,3-*cd*]pyrene
Phenylthiocarbamide	1-Phenyl-2-thio urea
TDE	1,1-Dichloro-2,2-bis (*p*-chlorophenyl)ethane (DDD)
Tetrachloromethane	Carbontetrachloride

Subject index

Acceptability rating	30, 85
Acetophenon	105
Actinomycetes	106
Activated carbon filtration	108
Adaptation	7, 10, 55
Additivity	8
ADI	86
Age, distribution of	23
Air	
- contamination	30, 73, 87
- odour of ambient	31
Aldrin	86
Algae	14
Alkylbenzenes	102
Amphetamine	88
Amsterdam	3, 11, 19
Aniline	107
Anilines, chlorinated	15, 102
Anosmia	6, 35
Artificial recharge	107
Bath	26
Benzene	90
Benzenes, chlorinated	15, 76, 105
3,4-Benzopyrene	68, 89, 105
Biphenyl	78
Bis(2-chloroethyl) ether	89
Bis(2-chloroisopropyl) ether	70, 73 a.f., 102, 105, 109
Bis(dichloropropyl) ether	76
Bis(2-methoxyethyl) ether	76
Bitter	9 a.f.
Blue-green algae	14, 106
Bromochloroiodomethane	75
Bromoform	70, 73 a.f., 90
n-Butylmercaptan	105
n-tert.Butylphenol	105
Caffeine	11
Calcium	24, 27, 56
- chloride	57
- hydrocarbonate	58
- sulphate	59
Carcinogens	89 a.f.
Chlordane	87 a.f.
Chlorides	15, 57
Chlorination	45, 74, 75
Chlorine	15, 16, 73
- dissolved organic	109
- like taste	28, 44, 49, 79
Chloroaniline	76, 79, 105
2-Chloroethyl-4-nitrophenylsulfone	76
Chloroform	15, 72 a.f., 89 a.f., 102
Chlorophenol	14, 88, 105, 107
Chlorophyl	106
Chlorotoluidine	76, 79, 105
Cholera	111
Coffee	85
Colour	13, 24
- assessment	23, 24, 84
Common chemical sense	3
Compensation	8
Consumer	101, 110
Consumption	
- of water	39, 84 a.f.
- of tea	84
Copper	10, 13, 16
Cresol	105
Cross-modality matching	8, 96
Cyanophyceae	106
Cyclohexenone	73
DDT	67
Decanal	75, 78, 87
Decanes	72, 74
Dichloroacetone	73
Dichloroaniline	79, 105
Dichlorobenzene	14, 35, 38, 70, 76, 78, 88, 102, 105, 107
Dichlorophenol	105
1,2-Dichloropropane	76
Dieldrin	67, 86
1,1-Dimethoxyisobutane	72, 74
1,1-Dimethoxypropane	76
Dimethyldisulphide	15
Diphenyloxide	105
Discrimination threshold	7
Disinfectant-like smell	35
Dodecenone	75
Dordrecht	106, 108
Drug reactivity	11
Earthy	12, 15, 28, 35, 44, 49
Educational standard, distribution of	23
Effluents	110
Esthers, chlorinated	76
Ethanethiol	105
Ethylacrylate	105
Ethylbenzene	72, 74
Ethyl isothiocyanate	75
European communities	16, 109
Eutrophication	110
Fatigue	50
Fechner's law	7
Fish taste	105

Flats	15	Magnesium	11, 27, 56
Fluoranthene	60, 72	- chloride	56, 57
Forced choice method	6, 36, 97	- hydrocarbonate	56, 58
		- sulphate	59, 76
Gallup Poll	19	Manganese	13, 24
Gas stripping	68	Menstrual period	10
Gemnemic acid	11	Mesityloxide	106
Geosmin	14, 75, 78	Metallic taste	28, 44, 49, 63
Grey list	109	Method of scaling	96
Guaiacol	105	2-Methylbutanenitrile	75
Gustation	9	2-Methylbutyric acid	15
		2-Methyl, 6-chlorophenol	105
Habituation	50	2-Methylisoborneol	14, 75, 78
Haloforms	75, 89, 102	Methyl isobutyrate	72
Hardness		Methyl 2-methylbutyrate	74
- colour assessment and	25	α-Methylstyrene	105
- distribution of	22	2-Methylthiobenzthiazole	14
- occurrence of particles and	24	Metropolis Water Act	111
		Meuse	12, 14, 45
- skin formation on tea and	27	Miracle fruit	11
		Mississippi river	89
- taste assessment and	27	Musty	12, 25, 35
- tea taste and	27	Mutagens	89 a.f.
- visual quality and	24		
Head space analysis	68	Naphthalene	14, 72, 78, 105, 106
Hedonic attribute	9		
Heptachlor	67	2-Naphtol	105
Heptanal	75, 78	Nicotine	11, 90
Heptanone	75, 78	Nitrobenzene	107
Heptenol	75	Nonanal	74, 78
Hexachlorobenzene	67	Nonanol	74, 75
Hexachlorobutadiene	70, 78, 105, 107	Nonene	75
Hexachlorocyclohexane (see also lindane)	67, 78, 105	Nonenol	75
		Nutrients	107, 110
Hexachloroethane	75		
Hexanal	75	Octanal	75, 78
Hexyl butyrate	73, 76	Octanols	73, 75, 78
Houses, distribution of types of	21, 22	Octene	75, 78,
		Odorogram	71
Humic acids	11, 13, 74	Odorous compounds	14, 77 a.f.
Hydrocarbonates	15, 64	Odour	
Hydrochloric	10	- detection threshold	5
Hydrogen ions	11	- free glassware	96
Hydrogen sulphide	11, 14	- intensity	7, 99
Hyoscine butylbromide	11	- number	8, 94 a.f.
		- number, accuracy of	98
Indene	14, 106	- number, changes during treatment	106, 108
Inquiry	18, 84		
Iodine	11, 74	- rating	26, 28, 46, 84 a.f.
ß-Ionene	88	- seasonal changes in	14
Iron	10, 11, 13, 16, 24	- sensitivity, age and	36, 41, 77
Isoborneol	35, 38, 78	- sensitivity, domicile and	36
		- sensitivity, intra-individual changes	39
Just-noticeable-difference	7		
		- sensitivity, sex and	36, 40
LD_{50}	86, 146 a.f.	- sensitivity, source of drinking water and	41
Lead salts	11		
Lemonade	84	- sensitivity, smoking and	40
Lindane	86, 90, 107	- sensitivity test	36
London	19, 111	- source and	26

- Threshold Concentration (OTC)	5, 77, 105, 146 a.f.
- quality	8, 28
Olfaction	4
Olfactory epithelium	3
Oral cavity	3
Ozone	107
Ozonization	45, 107
Panels	41, 52, 101
Particles, brown or black	24
Pesticides	86
Phenol	15
Phenols, chlorinated	12, 15
Phenylthiocarbamide	10
Phthalates	72, 74, 89, 102
Pilot inquiry	20
Polynuclear aromatic hydrocarbons	67, 89, 102
Potassium fluoracetate	1
6-n-Propylthiouracil	10
Psychophysics	4
Putrid	28, 44, 49
Pyridine	40
Quinine	10, 88
- sulphate	10
4-Quinone	105
- Resins, ion-exchange	109
Reuse	109, 110
Rhine	12, 45, 88, 108, 109
- convention against chemical pollution	16, 109
Rotterdam	12
Safety rating	32, 85
Saliva	9, 10, 55, 61
Salty	9, 11
Sampling of tapwater	44
Sensory interaction	11
Sesquiterpenes	15
Sex, distribution of	23
Shower	26
Signal detection theory	5
Smoking, effect on odour sensitivity	40
- effect on taste sensitivity	10
Sodium	
- carbonate	15
- chloride	10, 11, 56, 57
- hydrocarbonate	14, 56, 58
- sulphate	56, 59
Soup	85
Sour	9, 11
Source	
- distribution of	22
- knowledge of	29
Stains	13

Standards for drinking water	2, 62, 83, 99, 110
Stevens power law	7, 95
Storage reservoirs	75, 106, 107
Streptomycetes	14, 106
Structuurschema-1972	12
Styrene	105
Sucrose	10
Sulphates	15, 59
Sweet	9, 11, 89
Synergy	8
Tartaric acid	10
Taste	
- buds	9
- intensity	10
- intensity estimation by measuring chemicals	102
- Number	107, 108
- Number and ozonation	108
- opinion on	29
- quality	11, 28
- rating	26, 28, 46, 84, 102, 108
- rating, odour sensitivity and	50
- rating, sex and	49
- scale	43
-, source and	26
- Threshold Concentration (TTC)	55, 77
-, year of construction of house and	28
Tea flavour	26, 28, 85
Temperature	
- assessment	25, 84
-, effect on taste intensity	10
-, source	25
-, type of house and	25
Teratogens	89 a.f.
Terpenes	106
Tetrachloromethane	72, 74, 75, 89
Tetrachloroethene	70, 74, 107
Thames river	111
Toluene	70, 72, 74
Toxicity	
-, acute	87, 93
-, chronic	86, 93
Triads, method of	42
Trichlorobenzene	76, 78, 105
Trichloroethene	73, 89
Trichloronitromethane	73 a.f.
Trichlorophenol	105
Triethyl phosphate	76
Trifluoperazine	11
1,3,5-Trimethylbenzene	14, 70, 78, 105
Turbidity	13
- assessment	23

Undecanal	75	Weber's law	7
Undecene	75	WHO	99
US EPA	100, 104		
		XAD	69
Vecht	1	Xylenes	40, 72
VEWIN	19, 20		
		Zinc	13, 16
Warning mechanism	11, 83 a.f., 93		
Waste water	109		

References

Amoore, J.E., 1952, The stereochemical specificities of human olfactory receptors, *Perfum. Essent. Oil Rec., 43,* 321
Amoore, J.E., Venstrom, D., 1966, Sensory analysis of odor qualities in terms of the stereochemical theory, *J. of Food Sci, 31,* 118
Amoore, J.E., Venstrom, D., 1967, Specific anosmia: a clue to the olfactory code, *Nature, 214,* 1095
Amoore, J.E., Venstrom, E., Davis, A.R., 1968, Measurement of specific anosmia, *Perceptual and Motor Skills, 26,* 143
Anderson, C.T., Maier, W.J., 1979, Trace organics removal by anion exchange resins, *J.Am.Water Wks.Ass., 72,* 278
Anon., 1977, De Duitsers over hun drinkwater, H_2O, *10,* N43
Arfman, B.L., Chapanis, N.P., 1962, The relative sensitivities of taste and smell in smokers and non-smokers, *J. Gen. Psych., 66,* 315
Bacon, Sir Francis, 1627, Sylva Sylvarum, or a natural history in ten centuries, In: *Laus Aquae* (Ed. Leeflang, K.W.H.), Papyrus, Amsterdam, 1973
Baker, R.A., 1963, Threshold odors of organic chemicals, *J. Am. Water Wks Ass., 55,* 913
Bartoshuk, L.M., Dateo, G.A., Vandenbelt, D.J., Buttrick, R.I., Long, L.Jr., 1969, Effects of Gymnema sylvestre and Synsepalum dulcifirum on taste in man. In: *Olfaction and Taste III* (Ed. Pfaffman, C.M.), Rockefeller, New York, 436
Bartoshuk, L.M., 1974, NaCl thresholds in man: thresholds for water taste or NaCl taste? *Journ. of Comparative and Physiological Psychology, 87,* 2, 310
Bays, L.R., Burman, N.P., Lewis, W.M., 1970, Taste and odour in water supplies in Great Brittain: a survey of the present position and problems for the future, *Water Treatm. Exam., 2,* 136
Beets, M.G.J., 1973, Structure-response relationships in chemoreception, In: *Structure-Activity Relationships* (Amsterdam, P.O. Box 7938), 1, 225
Beidler, L.M., 1964, Taste receptor stimulation, *Biophysical Chemistry, 12,* 107
Bellar, T.A., Lichtenberg, J.J., Kroner, R.C., 1974, The occurrence of organohalides in chlorinated drinking water, *J. Am. Water Wks Ass., 66,* 703
Biemond, C., 1940, *Rapport 1940 inzake de watervoorziening van Amsterdam,* 105
Boelens, H., 1976, Molecular structure and olfactive properties, In: *Structure-Activity relationships in chemoreception* (Ed. Benz, G.), Information Retrieval Limited, London, 197
Boorsma, H.J., Zoeteman, B.C.J., Kraayeveld, A.J.A., 1969, Identificatie van reuk- en smaakstoffen in grondwater na selectieve ophoping, H_2O, *2,* 326
Borneff, J., Kunte, H., 1969, Carcinogenic substances in water and soil, Part XXVI: Routine method for the determination of polycyclic aromates in water, *Arch. Hyg., 153,* 220
Brown, K.S., MacLean, C.M., Robinette, R.R., 1968, The distribution of the sensitivity to chemical odors in man, *Human Biology, 40,* 456
Bruvold, W.H., Pangborn, R.M., 1966, Rated acceptability of mineral taste in water, *Journ. of Appl. Ps., 50,* 22
Bruvold, W.H., Gaffey, W.R., 1969, Evaluative ratings of taste in water, *Perceptual and Motor Skills, 28,* 179
Bruvold, W.H., Ongerth, H.J., 1969, Taste quality of mineralized water, *J. Am. Water Wks Ass., 61,* 170
Bruvold, W.H., Ongerth, H.J., Dillehay, R.C., 1969, Consumer Assessment of mineral taste in domestic water, *J. Am. Water Wks. Ass., 61,* 575
Bruvold, W.H., 1970, Laboratory panel estimation of consumer assessments of taste and flavor, *J. Appl. Psychol., 54,* 326
Bruvold, W.H., 1977, Consumer attitudes towards taste and odor in water, *J.Am.Water Wks Ass., 69,* 562
Bryan, P.E., Kuzminski, L.N., Sawyer, F.M., Feng, T.H., 1973, Taste thresholds of halogens in water, *J. Am. Water Wks Ass., 65,* 363
Burttschell, R.M., Rosen, A.A., Middleton, F.M., Ettinger, M.B., 1959, Chlorine derivatives of phenol causing taste and odor, *J. Am. Water Wks Ass., 51,* 205
Buswell, A.M., 1928. The chemistry of water and sewage treatment, *American chemical society monograph series, no. 38,* Chemical catalog company, Inc., New York, 104
Buyzer, F.C.G. de, 1974, De consument en de waterleiding, H_2O, *7,* 256

Cain, W.S., Moskowitz, H.R., 1974, Psychophysical scaling of odor, In: *Human responses to environmental odors,* (Ed. Turk, A., Johnston, Jr. J.W., Moulton, D.G.), Academic Press, New York and London, 1.

Cain, W.S., 1977, Differential sensitivity for smell: "noise" at the nose, *Science, 195,* 796

Centraal Bureau voor de Statistiek (CBS), 1976, *Bevolking der gemeenten van Nederland op 1 januari 1976,* Staatsuitgeverij 's-Gravenhage

Christensen, H.E., Luginbyhl, T.T., Carroll, B.S., 1974, *The toxic substances list, 1974 Edition,* U.S. Department of Health, Education and Welfare, National Institute for Occupational Safety and Health, Rockville, Maryland, 20852, U.S.A.

Cohen, J.M., Kamphake, L.J., Harris, E.K., Woodward, R.L., 1960, Taste threshold concentrations of metals in drinking water, *J. Am. Water Wks Ass., 52,* 660

Commission of the European Communities, 1975, Proposal for a directive of the Council concerning the quality of water destined for human consumption, *Publ. Bull. of the Eur. Comm., Nr. C 214,* 6

Coombs, C.H., 1964, A theory of data, Wiley, New York

Council of the European Communities, 1975, Directive of the Council of 16 June 1975 concerning the required quality of surface water intended for the production of drinking water in the Member states, *Publ. Bull. of the Eur. Comm., Nr. L 194,* 38

Council of the European Communities, 1976, Directive of the Council of 4 May 1976 concerning the contamination caused by certain dangerous substances which are discharged in the aquatic environment of the Community, *Publ. Bull. of the Eur. Comm., Nr. L 129,* 28

Crocker, E.C., Henderson, L.F., 1927, Analysis and classification of odours. An attempt to develop a workable model, *Amer. Perfum., 22,* 325, 356

Crompton, 1873/1874, *Bgham Med. Rev., 2* and *3*

Davies, J.T., Taylor, F.H., 1959, The role of absorption and molecular morphology in olfaction: the calculation of olfactory thresholds, *Biol. Bulletin, 117,* 2, 222

Dodd, G.H., 1974, Structure and function of chemoreceptor membranes, In: *Transduction mechanisms in chemoreception,* (Ed. Poynder, T.M.), Information Retrieval Ltd., London, 103

Dravnieks, A., 1972, Odor perception and odorous air pollution, *Tappi, 55,* 5, 737

Drinkwaterleiding Rotterdam, 1921-1967, Yearly Reports

Drost, G., 1971, Over (zout)smaak valt te twisten, H_2O, *4,* 432

Drost, G., Zoeteman, B.C.J., 1976, Zintuiglijk waarneembare aspecten van de waterkwaliteit, In: *Gezond Drinkwater,* (Ed., Zoeteman, B.C.J.) Staatsuitgeverij, 's-Gravenhage

Dutler, H., 1976, Structure-activity relationship as studied with an enzyme of know structure: the liver-alcohol-dehydrogenase-catalyzed reduction of alicyclic ketones, In: *Structure-activity relationships in chemoreception,* (Ed. Benz, G.), Information Retrieval, Ltd., London, 65

Engen, T., 1971, Psychophysics, In: *Experimental Psychology,* (Ed., Kling, J.W. and Riggs, L.A.), Holt, Rinehart and Winston, New York

Engen, T., 1974, Method and theory in the study of odor preferences, In: *Human responses to environmental odors,* (Ed. Turk, A., Johnston, Jr. J.W., Moulton, D.G.), Academic Press, New York and London, 122

English, J.N., Bennett, E.R., Lindstedt, K.D., 1977, Research required to establish confidence in the potable reuse of waste water, *J.Am. Water Wks Ass., 69,* 13

Fechner, G., 1859, *Elemente der Psychophysik,* Leipzig

Ferguson, D.B., 1975, Salivary Glands and Saliva, In: *Applied Physiology of the mouth,* (Ed. Lavelle, G.L.B.) Wright, Bristol, 145

Fischer, R., Griffin, F. (1964) Pharmacogenic aspects of gustation, *Arzneim. Forschung, 14,* 673

Fischer, R., Griffin, F., Archer, R.C., Zinsmeister, S.C., Jastram, P.S., 1965, Weber ratio in gustatory chemo reception; an indicator of systematic (drug) reactivity, *Nature, 207,* 1049

Fischer, R., 1971, Gustatory, behavioral and pharmacological manifestations of chemoreception in man, In: *Gustation and olfaction,* (Ed. Ohloff, G. and Thomas, A.F.), Academic Press, London, New York, 187

Foster, D., 1963, Odors in series and parallel, *Proceedings of Scientific Section, 39,* 1

Gallup Poll, 1973, Water quality and public opinion, *J. Am. Water Wks Ass., 65,* 513

Geldard, F.A., 1953, *The human senses,* John Wiley and Sons Inc., New York, 130

Gemert, L.J. van, Nettenbreijer, A.H., 1977, *Compilation of odour threshold values in air and water,* National Institute for Water Supply, P.O. Box 150, 2260 AD Leidschendam, The Netherlands, and Central Institute for Nutrition and Food Research TNO, Zeist, The Netherlands

Gerber, N.N., 1968, Geosmin from micro organisms is trans-1, 10-dimethyl, trans 9-decalol, *Tetrahedron Lett., 25,* 2971

Griffin, A.E., 1960, Significance and removal of manganese in water supplies, *J. Am. Water Wks Ass., 52*, 1326

Griffin, F., 1966, *On the interaction of chemical stimuli with taste receptors*, Dissertation, The Ohio State University, U.S.A.

Grob, K., Grob, G., Grob, K.Jr., 1975, Organic substances in potable water and its precursors, Part III: The closed-loop stripping procedure compared with rapid liquid extraction, *Journ. of Chrom., 106*, 299

Grob, K., Zürcher, F., 1976, Stripping of trace organic substances from water, Equipment and procedure, *Journ. of Chrom., 117*, 285

Grunt, F.E. de, Zoeteman, B.C.J., Piet, G.J., Heuvel, M.P.H. van der, 1977, Methods for the sensory evaluation of water quality, *RID-Mededeling*, 77-5, P.O. Box 150, 2260 AD Leidschendam, The Netherlands

Hall, E.S., Packham, R.F., 1965, Coagulation of organic color with hydrolyzing coagulants, *J. Am. Water Wks Ass., 57*, 1149

Hansen, R.E., 1972, Granular activated carbon filters for taste and odor removal, *J.Am. Water Wks Ass., 64*, 176

Harmsen, K., Bührer, H., Wieczorek, H., 1976, a-Glucosidases as sugar receptor proteins in flies, In: *Structure-activity relationships in chemoreception*, (Ed. Benz, G.), Information Retrieval Ltd., London, 79

Harper, R., 1972, *Human senses in action*, Churchill, Livingstone, 239

Henion, K.E., 1971, Olfactory intensity of diluted n-aliphatic alcohols, *Psychonomic Science, 22*, 213

Heymann, J.A., 1931-1932, De Rijn als toekomstige bron der drinkwatervoorziening van Amsterdam, *Water, 15*, 118, 131, 141, *Water, 16*, 219, 227

Hoogdrukwaterleiding van de gemeente Zwolle, 1908, Yearly Report 1907, De Erven, Tijl, Zwolle, 12

Howard, C.D., 1923, Zinc contamination in drinking water, *J. Am. Water Wks. Ass., 10*, 411

Hrubec, J., 1978, Resultaten van de proeven met de infiltratie van voorgezuiverd Rijnwater in Veluwezand, *RID-Mededeling* 78-1, P.O. Box 150, 2260 AD Leidschendam, The Netherlands

Hughes, G., 1969, Changes in taste sensitivity with advancing age, *Geront. clin., 11*, 224

International Union of Pure and Applied Chemistry (IUPAC), 1969, *Nomenclature of Organic Chemistry*, Butterworths, London

Joyce, C.R.B., Pan, L., Varonos, D.D., 1968, Taste sensitivity may be used to predict pharmacological effects, *Life Sciences, 7*, I, 533

Junk, G.A., Richard, J.J., Grieser, M.D., Witiak, D., Witiak, J.L., Guello, M.D., Vic, R., Svec, H.J., Fritz, J.S., Calder, G.C., 1974, Use of macro-reticular resins in the analysis of water for trace organic contaminants, *J. Chrom., 99*, 745

Junk, G.A., Chriswell, C.D., Chang, R.C., Kissinger, L.D., Richard, J.J., Fritz, J.S., Svec, H.J., 1976, Applications of resins for extracting organic components from water, *Z. Anal. Chem., 282*, 331

Kahn, R.J., 1965, Sensory threshold fluctuations as a function of menstrual cycle phase, *Diss. abstr., 26*, 6, 3478

Kaplan, A.R., Glanville, E.V., Fischer, R., 1964, Taste threshold for bitterness and cigarette smoking, *Nature, 202*, No. 4939, 1366

Kaplan, A.R., Fischer, R., 1965, Taste sensitivity for bitterness, some biological and clinical implications, In: *Recent advances in Biological Psychiatry, Vol. III*, (Ed. Wortis, J.), Plenum Press, New York. 183

Kaplan, A., Glanville, E., Fischer, R., 1965, Cumulative effect of age and smoking on taste sensitivity in males and females, *J. Geront., 20*, 334

Kehoe, R.A., Cholak, J., Largent, E.J., 1944, The hygienic significance of the contamination of water with certain mineral constituents, *J. Am. Water Wks Ass., 36*, 645

Kijima, H., 1976, Specificity of two receptor sites and identification of receptor in sugar receptor of flies, In: *Structure-activity relationship in chemoreception*, (Ed. Benz, G) Information Retrieval Ltd., London, 89

Knapp, W., Fischer, R., Beck, J., Teitelbaum, A., 1966, *Dis. nerv. Syst., 27*, 729

Kniebes, D.V., Chrisholm, J.A., Stubbs, R.C., 1969, Odors and odorants, In: *The Engineering View*, ASHRAE, New York, 27

Koelega, H.S., Köster, E.P., 1974, Some experiments on sex differences in odor perception, *Ann. N.Y. Acad. Sci., 237*, 234

Kölle, W., Koppe, P., Sontheimer, H., 1970, Taste and odour problems with the river Rhine, *Water Treatm. Exam., 2*, 120

Kölle, W., Schweer, K.H., Güsten, H., Stieglitz, L. 1972, Identifizierung schwer abbaubaren Schadstoffen im Rhein und Rheinuferfiltrat, *Vom Wasser, 39,* 109

Koninkrijk der Nederlanden, 1977, Overeenkomst inzake de bescherming van de Rijn tegen chemische verontreiniging, met bijlagen; Bonn, 3 december 1977, *Tractatenblad, 32,* 26

Koppe, P., 1965, Identifizierung der Hauptgeruchsstoffe im Uferfiltrat des Mittel- und Niederrheins, *Vom Wasser, 32,* 33

Köster, E.P., 1969, Intensity in mixtures of odorous substances, In: *Olfaction and taste III,* (Ed. Pfaffmann, C.M.) Rockefeller, New York, 142

Köster, E.P., 1971, *Adaptation and cross-adaptation in olfaction,* Thesis, University of Utrecht, The Netherlands

Köster, E.P., 1975, Human psychophysics in olfaction, In: *Methods in olfactory research,* (Ed., Moulton, D.G., Turk, A., Johnston, Jr. J.W.) Academic Press, London, 345

Kraybill, H.F., 1976, Global distribution of carcinogenic pollutants in water, *Ann. N.Y. Acad. Sci., 298,* 80

Kraybill, H.F., Tucker Helmes, C., Sigman, C., 1978, Biomedical aspects of biorefractories in water, In: **Aquatic Pollutants (Ed. Hutzinger, O, Lelyveld, H. van, Zoeteman, B.C.J.), Pergamon Press, Oxford, 419**

Kroeze, J., 1971, University of Utrecht, Psychological Laboratory, Varkenmarkt 2, Utrecht, The Netherlands, unpublished data

Kruithof, J.C., 1978, *Zuivering van Rijnwater met behulp van coagulatie en ozonisatie,* Thesis Univ. of Delft, The Netherlands, 161

Kühn, W., Sontheimer, H., Kurz, R., 1978, Use of ozone and chlorine in waterworks in the Federal Republic of Germany, In: *Ozone/Chlorine dioxide oxidation products of organic materials* (Ed. Rice, R.G., Cotruvo, J.A.), Ozone Press Int., Cleveland, Ohio 441079

Laffort, P., 1969, A linear relationship between olfactory effectiveness and identified molecular characteristics, extended to fifty pure substances, In: *Olfaction and Taste III,* (Ed. Pfaffmann, C.M.), Rockefeller, New York, 151

Leeflang, K.W.H., 1974, *Ons drinkwater in de stroom van de tijd,* VEWIN, Rijswijk, 24

Leventer, H., Eren, J., 1969, Taste and odor in the reservoirs of the Israel national water system, In: *Developments in Water Quality Research,* Proc. Jerusalem int. conf. on water quality and pollution research

Lockhart, E.E., Tucker, C.L., Merritt, M.C., 1955, The effect of water impurities on the flavor of brewed coffee, *Food Research, 20,* 598

Maarse, H., Ten Oever de Brauw, M.C., 1972, *Chemisch Weekblad, 68,* 11

McBurney, D.H., 1969, Effects of adaptation on human taste function, In: *Olfaction and taste III,* (Ed. Pfaffmann, C.M.) Rockefeller, New York, 407

McCreary, J. J., Snoeyink, V.L., 1977, Granular activated carbon in water treatment, *J.Am. Water Wks Ass., 69,* 437

McKee, J.E., Wolf, H.W., 1963, *Water quality criteria,* (2nd edition) The resources agency of California, State Water Quality Control Board, Publication No. 3-A, Sacramento, California, USA, 95814

McNamara, B.P., Danker, W.H., 1968, Odor and taste, In: *Basic principles of sensory evaluation,* ASTM STP 433, American society for testing and materials, Philadelphia, Pa., USA, 13

Medsker, L.L., Jenkins, D., Thomas, J.F., 1968, Odorous compounds in natural waters. An earthy smelling compound associated with blue-green algae and actinomycetes, *Env. Sci. Techn., 2,* 461

Mefferd, R.B., Wieland, B.A., 1968, Taste thresholds for sodium chloride in longitudinal experiments, In: *Perceptual and motor skills, 27,* 295

Meijers, A.P., 1970, *Onderzoek naar organische stoffen in rivierwater en drinkwater,* Dissertation, University of Delft, The Netherlands, 152

Meijers, A.P., 1977, Ozonisatie, $H_2O, 18,$ 417

Meiselman, H.L., 1972, Human taste perception, *CRC Critical Reviews in Food Technology,* 89

Meiselman, H.L., Halpern, B.P., 1973, Enhancement of taste intensity through pulsatile stimulation, *Physiol. and Behaviour, 11,* 713

Meiselman, H.L., 1976, Psychology of taste, *Cereal Foods World, 21,* 2, 52

Menco, B.P.M., Dodd, G.H., Davey, M., Bannister, L.H., 1976, Presence of mebrane particles in freeze-etched bovine olfactory cilia, *Nature, 263,* 597

Menco, B.P.M., 1977, *A qualitative and quantitative investigation of olfactory and nasal respiratory mucosal surfaces of cow and sheep based on various ultrastructural and biochemical methods,* Thesis, Landbouwhogeschool, Wageningen, The Netherlands, 114

Menevse, A., Dodd, G., Poynder, T.M., 1977, Evidence for the specific involvement of cyclic AMP in the olfactory transduction mechanism, *Biochem. Biophys. Research Comm.*, 77, 2, 671

Mitchell, M.J., 1967, *Some interrelationships of the chemical senses, smell, and irritance, at the near-threshold level, and trends within the series of n-aliphatic monohydric alcohols*, M.S. Thesis, University of Canterbury, Christchurch, New Zealand

Moncrieff, R.W., 1966, *Odour preferences*, Wiley, New York

Moncrieff, R.W., 1967, *The chemical senses*, Leonard Hill, London

Morra, C.F.H., Linders, J.B.H.J., Den Boer, A.C., Ruygrok, C.Th.M., Zoeteman, B.C.J., 1979, Organic compounds measured during 1978 in the river Rhine in The Netherlands, *RID-Mededeling* 79-3, P.O. Box 150, 2260 AD Leidschendam, The Netherlands

Moskowitz, H.R., 1970a, Sweetness and intensity of artificial sweeteners, *Percept. Psychophys.*, 8, 40

Moskowitz, H.R., 1970b, Taste intensity as a function of stimulus concentration and solvent viscosity, *Journ. of texture studies*, 1, 502

Moskowitz, H.R., 1971, Ratio scales of acid sourness, *Percept. Psychophys.*, 9, 371

Moskowitz, H.R., 1973, Effects of solution temperature on taste intensity in humans, *Physiology and behaviour*, 10, 289

Moulton, D.G., 1971, Detection and recognition of odor molecules, In: *Gustation and olfaction*, (Ed. Ohloff, G. and Thomas, A.F.), Academic Press, London and New York

National Academy of Sciences, 1977, *Drinking Water and Health*, Washington D.C., 20418, U.S.A., 489

O'Mahony, M., 1972, Salt taste sensitivity; a signal detection approach *Perception*, 1, 459

Oskam, G., Rook, J.J., 1970, Nogmaals: reuk en smaak van rivierwater i.v.m. voorraadvorming, H_2O, 3, 216

Packham, R.F., 1968, *Taste and odour in water supply*, T.M. 48 Water Research Association, Medmenham, Marlow, Bucks., U.K.

Page, G.G., 1973, Contamination of drinking water by corrosion of copper tubes, *New Zealand Journ. of Science*, 16, 349

Page, T., Harris, R.H., Epstein, S.S., 1976, Drinking water and cancer mortality in Louisiana, *Science*, 2 July, 55

Pangborn, R.M., Trabue, I.M., Little, A.C., 1971, Analysis of coffee, tea and artificially flavored drinks prepared from mineralized waters, *J. Food Sci.*, 36, 355

Pangborn, R.M., Bertolero, L.L., 1972, Influence of temperature on taste intensity and degree of liking of drinking water, *J. Am. Water Wks Ass.*, 64, 511

Pettenkofer, Max von, 1873, *Was man gegen die Cholera tun kann*, Ansprache an das Publikum, Oldenbourg, München, 39

Piet, G.J., Zoeteman, B.C.J., Kraayeveld, A.J.A., 1972, Earthy smelling substances in surface waters of The Netherlands, *Water Treatm. Exam.*, 21, 4, 281

Piet, G.J., Zoeteman, B.C.J., Klomp, R., 1975, Polynuclear aromatic hydrocarbons in the water environment of The Netherlands, *RID-communication* 75-6 (P.O. Box 150, 2260 AD Leidschendam, The Netherlands)

Rand, M.C., Greenberg, A.E., Taras, M.J., Franson, M.A., 1975, *Standard Methods for the Examination of Water and Wastewater* (14th edition) Am. Publ. Health Ass., Washington D.C. 20036, 75, 121

Richter, C.P., 1950, *J. Comp. Physiol. Psychol.*, 43, 358

Rook, J.J., 1974, Formation of haloforms during chlorination of natural waters, *Water Treatm. Exam.*, 23, 234

Rook, J.J., 1975, Bromierung organischer Wasserinhaltstoffe als Nebenreaktion der Chlorung, *Vom Wasser*, 44, 57

Rosen, A.A., Skeel, R.T., Ettinger, M.B., 1963, Relationship of river water odor to specific organic contaminants, *J. Water Poll. Contr. Fed.*, 35, 6, 777

Rosen, A.A., Mashni, C.I., Safferman, R.S., 1970, Recent developments in the chemistry of odour in water: the cause of earthy/musty odour, *Water Treatm. Exam.*, 19, 106

Roskam, E., 1970, *Nonmetric multidimensional scaling "minissa-1", version for triadic data*, University of Nijmegen, Psychological Department, Nijmegen, The Netherlands

Ryazanov, V.A., 1962, Sensory physiology as basis for air quality standards (the approach used in the Soviet Union), *Archives of Env. Health*, 5, 480

Safferman, R.S., Rosen, A.A., Mashni, C.I., Morris, M.E., 1967, Earthy smelling substances from blue-green alga, *Env. Sci. Techn.*, 2, 461

Shapiro, J., 1964, Effect of yellow organic acids on iron and other metals in water, *J. Am. Water Wks Ass.*, 56, 1062
Shuval, H.I., 1977, *Water renovation and reuse*, Acad.Press, New York
Siegel, S., 1956, *Non parametric statistics*, Int. Student Ed., McGraw-Hill, Kogakusha, Ltd., Tokyo, 184
Silvey, J.K., Henley, O.E., Wyatt, J.T., 1972, Growth and odor-production studies, *J. Am. Water Wks Ass.*, 64, 35
Soltan, H.C., Bracken, S.E., 1958, The relation of sex to taste reactions, *The Journ. of Heredity*, 44, 280
Stein, M., Ottenberg, R. and Roulet, N., 1958, A study of the development of olfactory preferences, *American Med. Ass., Archives of Neurological Psychiatry*, 80, 264
Stevens, S.S., 1957, *Psychol. Rev.*, 64, 153
Stone, H., Pangborn, R.M., 1968, Intercorrelation of the senses, In: *Basic principles of sensory evaluation*, ASTM STP 433, American society for testing and materials, Philadelphia, Pa, USA, 30
Stone, H., Oliver, S., Kloehn, J., 1969, Temperature and pH effects on the relative sweetness of supra-threshold mixtures of dextrose and fructose, *Percept. Phychophys.*, 5, 257
Structuurschema Drink- en Industriewatervoorziening-1972, 1975, Deel c: regeringsbeslissing, *Tweede Kamer*, zitting 1974-1975, 13 337, nrs. 4-5, Staatsdrukkerij en Uitgeverijbedrijf, 's-Gravenhage, 25
Stuiver, M., 1958, *Biophysics of the sense of smell*, Thesis, University of Groningen, The Netherlands
Summer, W., 1970, *Geruchlosmachung von Luft und Abwasser*, R. Oldenburg, München und Wien, 197
Swets, J.A., Tanner, W.P. and Birdsall, T.G., 1961, Decision processes in perception, *Psychol. Rev.*, 68, 301
Teranishi, R., 1971, Odor and molecular structure, In: *Gustation and olfaction*, (Ed. Ohloff, G. and Thomas, A.F.), Academic Press, London and New York, 165
Tissandier, G., 1873, *L'eau*, Librairie Hachette et Cie, Paris (3rd edition), 288
Torgerson, W.S., 1967, *Theory and methods of scaling*, (7th edition) Wiley, New York
U.S. Environmental Protection Agency, 1973, *Water Quality Criteria 1972*, EPA.R3.73.0.33, U.S. Government Printing Office, Stock Nr. 5501-00520, Washington D.C., 74
U.S. Environmental Protection Agency, 1975, *Preliminary assessment of suspected carcinogens in drinking water*, Report to Congress by Office of Toxic Substances, EPA, Washington D.C. 20460
U.S. Environmental Protection Agency, 1977, Drinking water and health recommendations of the National Academy of Sciences, *Federal Register*, 42, No. 132, Monday, July 11, Part. III, USA
VEWIN, 1976, Lijst van waterhardheid per gemeente in Nederland, *VEWIN Jaarboek 1976*, Postbus 70, Rijswijk, 186
Vlugt, J.C. van der, Zoeteman, B.C.J., Piet, G.J., Schippers, J.C., Burg L. van de, 1973, Plankton en reukstoffen in het spaarbekken "De Grote Rug", in 1971, H_2O, 439
Wegman, R.C.C., Van den Broek, H.H., Hofstee, A.W.M., 1978, Chloorphenolen in Nederlands oppervlaktewater, *RIV-rapport* nr. 154/78 RA, P.O. Box 1, Bilthoven, The Netherlands
Whipple, G.C., 1907, *The value of pure water*, New York, Wiley
Windle Taylor, E., 1965, Symposium on consumer complaints, *Water Treatment. Exam.*, 14, 2, 99
Winneke, G., Kastka, J., 1976, Zur Wirkung von Geruchsstoffen in Labor- und Felduntersuchungen, *Zbl. Bakt. Hyg. I. Abt. Orig.*, B 162, 41
Winneke, G., Kastka, J., 1977, Odor pollution and odor annoyance reactions in industrial areas of the Rhine-Ruhr Region, In: *Proceedings of the 6th International Symposium on Olfaction and Taste*, Information Retrieval Ltd., London
World Health Organization, 1971, *International standars for drinking water*, World Health Organization Geneva (3rd edition)
World Health Organization, 1973, Pesticide residues in food, Report of the 1972 joint FAO/WHO meeting, *WHO Technical Report Series, No. 525*, Geneva
World Health Organization, International Reference Centre for Community Water Supply, 1975, Health effects relating to direct and indirect re-use of waste water for human consumption, *WHO-IRC Technical Paper Series, No. 7*, P.O. Box 140, 2260 AD Leidschendam, The Netherlands, 39
World Health Organization, 1976, Pesticide residues in food, Report of the 1975 joint FAO/WHO meeting, *WHO Technical Report Series,, No. 592*, Geneva
Woskow, M.H., 1968, Multidimensional scaling of odours, In: *Theories of odour and odour measurement*, (Ed. Tanyolac, N.) Robert College, Bebek, Istanbul
Wright, R.H., Michels, K.M., 1974, Evaluation of far infrared relations to odor by a standard similarity method, *Ann. N.Y. Acad. Sci.*, 116, 535

Yoshida, M., 1964, Studies in psychometric classification of odors (4) (5), *The Jap. Psychol. Res.*, *6*, 115, 145
Zoeteman, B.C.J., 1970a, Zuiveringsmethoden van zoet grondwater voor de drinkwatervoorziening in Nederland, H_2O, *3*, 44
Zoeteman, B.C.J., 1970b, Reuk en smaak van rivierwater in verband met voorraadvorming, H_2O, *3*, 136
Zoeteman, B.C.J., Kraayeveld, A.J.A., Piet, G.J., 1971, Oil pollution and drinking water odour, H_2O, *4*, 367
Zoeteman, B.C.J., Piet, G.J., 1972/1973, On the nature of odours in drinking water resources of The Netherlands, *Science Tot. Env.*, *1*, 399
Zoeteman, B.C.J., Piet, G.J., 1973, Drinkwater is nog geen water drinken, H_2O, *6*, 7, 174
Zoeteman, B.C.J., Piet, G.J., 1974a, Meting van reukstoffen in water, mogelijke oorzaken van hinderlijke watergeur, H_2O, *7*, 162
Zoeteman, B.C.J., Piet, G.J., 1974b, Cause and identification of taste and odour compounds in water, *Science Tot. Env.*, *3*, 103
Zoeteman, B.C.J., Piet, G.J., Morra, C.F.H., Grunt, F.E. de, 1975, Organic micropollutants in drinking water derived from surface water in The Netherlands, In: *Colloquium on Drinking Water Quality and Public Health*, (Ed. Packham, R.F.), Water Research Centre, High Wycombe, England
Zoeteman, B.C.J., Haring, B.J.A., 1976, *The nature and impact of deterioration of the quality of drinking water after treatment and prior to consumption*, Report to the Commission of the European Communities, Directorate of Social Affairs, Luxembourg (August), RID-rapport 76-14, P.O. Box 150, 2260 AD Leidschendam, The Netherlands
Zoeteman, B.C.J., Haring, B.J.A., Karres, J.J.C., 1976, Enquête naar drinkwater gebruiksgewoonten van de Nederlandse consument, H_2O, *20*, 410
Zoeteman, B.C.J., 1977, International cooperation in studying the health aspects of organic contaminants in indirectly reused waste water, *Ann. New York Acad.Sci.*, *298*, 561
Zwaardemaker, H., 1895, *Die Physiologie des Geruchs*, Engelmann, Leipzig
Zwaardemaker, H., 1907, Uber die Proportionen des Geruchs Kompensation, *Arch. Anat. Physiol.*, (Leipzig), *31*, Suppl. Bd., 59, 70

Appendix 2.1

Survey of z-values for associations between ordinal variables of the national inquiry

The standard normal deviate (z), which follows a standard normal distribution if the null-hypothesis is true, has been calculated by the CBS for the data obtained by means of the national inquiry (n = 3073) and for the variables presented in table 2.A.1. The null-hypothesis is rejected when $z \geqslant 1.64$ ($p \leqslant 0.05$).

Table 2.A.1: z-values for some associations between ordinal variables of the national inquiry

z-values	Colour rating	Odour rating	Taste rating	Tea flavour rating	Temperature rating	Acceptability rating	Safety rating	Hardness	Local air quality rating	Work air quality rating
Colour rating			12			16	13	5.0		
Odour rating			23			20	16	5.2		
Taste rating	12	23		21	11	35	17	11	12	5.2
Tea flavour rating			21			25		8.5		
Temperature rating		7	11			12	9			
Acceptability rating	16	20	35	25			20	11		
Safety rating	13	16	17		20			4.6		
Hardness	5.0	5.2	11	8.5		11	4.6			
Local air quality rating			12							
Work air quality rating			5.2							

In those cases where no z-values have been given in table 2.A.1, these have not been calculated.

Appendix 2.2

Survey of H-values for associations between ordinal and nominal variables of the national inquiry

The H-values, according to the Kruskal-Wallis test (Siegel, 1956) have been calculated by the CBS for the data of the national inquiry (n = 3073). The results of the calculated H-values are presented in table 2.A.2. The null-hypothesis is rejected e.g. for $H = 5.99$ ($\varphi = 2$, one tailed value of $p = 0.05$).

For the calculations 3 types of houses have been considered: *"one family houses"*, *"flats"* and *"others"*. In those cases where no H-values have been given in table 2.A.2, these have not been calculated.

Table 2.A.2: H-values for some associations between ordinal and nominal variables of the national inquiry

H-values	Source	Opinion on source	Sex	Type of house	Age of house	Domestic plumbing system	Brown and black particles	Skin on tea	Offensive skin on tea	Cause of bad smelling air	Education head household
Colour rating	470										
Odour rating	210		NS								
Taste rating	400	190	NS			NS	16				
Taste elsewhere rating	90										
Tea flavour rating	210	130	NS						65		
Temperature rating	50				130						
Acceptability rating	320										
Safety rating	120	70									
Hardness							7.9	270	NS		
Local air quality rating										33	NS

NS: not significant (p > 0.05)

Appendix 2.3

Some results of the national inquiry, arranged according to 50 communities, in combination with some water quality data

Community number	Number of observations	Water source	Average taste rating	Average taste elsewhere rating	Ca mg/l	Mg mg/l	Colour mgPt/l	Average colour rating	Average acceptability rating	Average safety rating	Average consumption of water as such l/day/person
1	2	3	4	5	6	7	8	9	10	11	12
1	30	GW	1.27	-0.55	20	5.0	1	1.10	1.40	1.00	0.205
2	28	GW	1.29	-0.74	31	1.6	2	1.00	1.25	1.00	0.295
3	42	GW	1.38	-0.67	21	2.7	2	1.10	1.48	1.02	0.505
4	30	GW	1.40	-0.65	36	2.3	4	1.27	1.47	1.07	0.405
5	28	GW	1.43	-0.50	45	2.9	7	1.14	1.61	1.00	0.430
6	29	GW	1.45	-0.47	41	3.3	12	1.10	1.97	1.03	0.200
7	40	GW	1.45	-0.35	110	10	2	1.03	1.45	1.05	0.220
8	31	GW	1.45	-0.41	30	3.8	3	1.10	2.00	1.16	0.315
9	42	GW	1.50	-0.65	71	7.0	10	1.00	1.43	1.00	0.390
10	29	GW	1.52	-0.61	26	1.8	1	1.03	1.48	1.00	0.290
11	32	GW	1.53	-0.44	39	3.1	12	1.25	1.72	1.13	0.420
12	42	GW	1.55	-0.24	12	1.7	2	1.17	1.86	1.05	0.480
13	33	GW	1.55	-0.44	42	3.3	2	1.24	1.64	1.21	0.370
14	35	GW	1.57	0.18	98	9.1	11	1.20	1.83	1.06	0.135
15	34	GW	1.62	-0.33	100	11	2	1.06	1.77	1.06	0.220
16	24	GW	1.63	-0.39	39	3.1	12	1.50	1.88	1.13	0.190
17	38	GW	1.63	-0.52	35	2.5	10	1.08	1.66	1.08	0.300
18	42	GW	1.71	-0.35	23	4.0	1	1.21	1.64	1.07	0.150
19	32	GW	1.72	-0.33	70	6.5	10	1.03	1.81	1.03	0.155
20	34	GW	1.74	-0.19	26	2.5	1	1.24	1.88	1.06	0.135
21	31	GW	1.74	-0.11	68	5.9	5	1.16	1.68	1.07	0.275
22	35	GW	1.74	-0.75	34	2.9	3	1.09	1.83	1.03	0.405

Community number	Number of observations	Water source	Average taste rating	Average taste elsewhere rating	Ca mg/l	Mg mg/l	Colour mgPt/l	Average colour rating	Average acceptability rating	Average safety rating	Average consumption of water as such 1/day/person
1	2	3	4	5	6	7	8	9	10	11	12
23	43	GW	1.74	-0.33	43	3.5	13	1.05	1.49	1.02	0.310
24	30	GW	1.77	-0.77	27	2.6	2	1.07	1.73	1.00	0.430
25	23	GW	1.78	-0.46	33	8.3	2	1.04	1.61	1.04	0.305
26	33	GW	1.82	-0.22	81	6.9	7	1.03	1.88	1.30	0.305
27	30	GW	1.83	-0.07	54	8.5	2	1.20	1.77	1.20	0.210
28	30	GW	1.83	-0.17	35	2.9	3	1.03	1.60	1.07	0.165
29	38	GW	1.87	-0.13	90	6.7	7	1.16	1.87	1.00	0.180
30	36	GW	1.92	-0.44	45	4.5	5	1.08	2.08	1.11	0.165
31	33	GW	1.94	-0.18	70	6.2	10	1.00	2.15	1.03	0.195
32	39	GW	1.95	0.08	39	3.1	12	1.28	1.88	1.08	0.185
33	27	GW	2.00	-0.50	17	1.3	2	1.15	1.70	1.15	0.200
34	35	GW	2.03	0.00	76	9.6	20	1.17	2.14	1.17	0.190
35	31	GW	2.03	-0.27	180	28	26	1.55	2.39	1.13	0.125
36	30	BF	2.07	-0.46	87	12	7	1.53	2.13	1.07	0.195
37	36	GW	2.19	0.12	160	12	5	1.50	2.86	1.47	0.260
38	122	SW	2.25	0.20	98	8.7	8	1.30	2.39	1.18	0.195
39	148	SW	2.25	0.02	90	9.0	12	1.24	2.35	1.23	0.225
40	39	SW	2.26	0.00	100	11	10	1.15	2.74	1.23	0.190
41	33	BF	2.39	-0.36	110	13	8	1.27	2.45	1.06	0.160
42	56	SW	2.48	-0.14	75	8.5	12	1.61	2.43	1.38	0.165
43	94	SW	2.65	0.50	51	9.2	5	1.44	2.72	1.40	0.225
44	28	BF	2.75	-0.15	90	12	7	1.25	2.93	1.32	0.195
45	23	SW	2.87	-0.17	110	12	20	1.48	2.83	1.35	0.135
46	27	SW	2.90	0.13	45	12	4	1.11	2.85	1.30	0.290
47	34	BF	2.97	0.46	73	9.0	10	1.44	3.32	1.38	0.250
48	30	SW	3.00	0.14	100	18	5	1.60	3.00	1.37	0.135
49	31	BF	3.07	0.47	87	12	7	1.13	3.16	1.13	0.145
50	33	BF	3.12	0.00	100	13	9	1.15	2.73	1.21	0.140

Explanation:

GW: ground water
BF : bank filtered water
SW : surface water

The chemical data are based on single measurements in 1975 by the National Institute for Public Health, Bilthoven.

The average ratings for each community have been calculated by considering each item on the category scales of equal weight. A total number of 1933 observations obtained by means of the inquiry have been used. The average ratings for the *"taste elsewhere"* item were calculated by rating the scores in the *"elsewhere better"* category as (-1), in the *"elsewhere worse"* category as (+1) and the *"elsewhere the same"* or *"elsewhere sometimes better, sometimes worse"* categories as (0).

Coefficients of correlation:

Several coefficients of correlation have been calculated by means of linear regression analysis for pairs of the parameters considered. The results of the relevant coefficients of correlation are presented in table 2.A.3. Where no values have been given these have not been calculated.

Table 2.A.3: Correlation coefficients (r) for several aspects of water quality assessment and some physical-chemical water quality data ($r \geq 0.28, \alpha \leq 0.05$)

Parameter	Taste elsewhere rating	Ca content	Mg Content	Colour intensity (mgPt/l)	Colour rating	Acceptability rating	Safety rating	Consumption of water as such
Taste rating	0.72	0.48	0.61		0.49	0.92	0.72	-0.47
Taste elsewhere rating						0.74	0.64	-0.51
Ca content			0.86	0.50	0.44	0.56	0.43	-0.48
Mg content								
Colour intensity (mgPt/l)					0.39			
Colour rating						0.56	0.63	-0.33
Acceptability rating							0.77	-0.44
Safety rating								-0.26

Appendix 3.1

Survey of the characteristic data of the group used for selecting the panel members and their odour sensitivity for aqueous solutions of isoborneol and o-dichlorobenzene

Explanation

Sex (X_1) : M=Male (0), F=Female (1)
Age (X_2) : 1=18 - 35 years (0)
  2=35 - 49 years (1)
  3=50 years and over (2)
Smoking (X_3) : —=no smoking (0)
  +=smoking 1 or more cigarettes per day
Source (X_4) : GW=ground water source (0)
  MIX=mixed ground water and surface water sources, including bank filtration and dune infiltration (1)
  SW=surface water source (2)
Community size (X_5) : 1=less than 10.000 inhabitants (0)
  2=10.000-75.000 inhabitants (1)
  3=more than 75.000 inhabitants (2)
ISO-selection : measured OTC of o-dichlorobenzene in water during the selection procedure at home
  1=0.3 μg/l; 2=1.5 μg/l;
  3=4.5 μg/l; 4=15 μg/l;

DCB-selection	$5=45\ \mu g/l$; $6=150\ \mu g/l$; $7=$ above $150\ \mu g/l$: measured OTC of o-dichlorobenzene in water during the selection procedure at home $1=3\ \mu g/l$; $2=15\ \mu g/l$; $3=45\ \mu g/l$; $4=150\ \mu g/l$; $5=450\ \mu g/l$; $6=1500\ \mu g/l$; $7=$ above $1500\ \mu g/l$
-In ISO-panel	: the negative natural logarithm of the calculated average OTC of isoborneol in water, using the data from the three panel sessions and the maximum likelihood method for calculation of the average value.
-In DCB-panel	: the negative natural logarithm of the calculated average OTC of o-dichlorobenzene in water, using the data from the three panel sessions and the maximum likelihood method for calculation of the average value

Calculated multiple linear regressions:

1. For the data obtained during the selection procedure (n=148)

   Threshold class $ISO = 3 + 0.49 X_1 + 0.26 X_2 + 0.40 X_3 + 0.33 X_4 - 0.03 X_5$

   Multiple coefficient of correlation: 0.289

   Threshold class $DCB = 4 - 0.19 X_1 + 0.11 X_2 + 0.47 X_3 - 0.16 X_4 + 0.21 X_5$

   Multiple coefficient of correlation: 0.194

2. For the data obtained during the panel sessions (n=54)

   $-\ln(ISO) = 3.56 - 0.43 X_1 - 0.17 X_2 + 0.19 X_3 - 0.45 X_4 + 0.05 X_5$

   Multiple coefficient of correlation: 0.463

   $-\ln(DCB) = 2.60 - 0.20 X_1 - 0.15 X_2 - 0.36 X_3 - 0.33 X_4 + 0.04 X_5$

   Multiple coefficient of correlation: 0.329

person nr.	sex	age	smoking	source	community size	ISO-selection	DCB-selection	-ln ISO-panel	-ln DCB-panel	person nr.	sex	age	smoking	source	community size	ISO-selection	DCB-selection	-ln ISO panel	-ln DCB-panel
1	F	3	+	GW	0	4	4			46	M	2	+	SW	2	5	4	2.67	0.43
2	F	1	—	GW	0	4	3	2.51	2.67	47	F	1	—	GW	1	4	2	4.10	2.18
3	M	1	+	SW	2	2	3	1.98	1.05	48	F	3	—	GW	0	3	3	4.71	3.46
4	M	1	+	SW	2	1	6	4.03	2.28	49	M	1	+	GW	0	1	5	3.62	3.05
5	F	1	+	GW	0	3	1	4.07	3.58	50	M	1	—	SW	2	1	1	2.64	0.33
6	F	1	+	GW	2	4	4	3.41	2:30	51	M	1	—	GW	1	3	5	4.08	3.92
7	M	1	+	SW	2	4	5			52	F	3	—	GW	1	2	2	3.62	3.05
8	F	3	—	SW	0	4	1	1.91	-0.31	53	M	1	—	MIX	2	2	2	4.19	2.34
9	F	1	+	MIX	2	4	3	2.12	1.12	54	F	2	+	SW	2	3	1	3.16	2.52
10	M	1	—	GW	1	3	3	3.50	1.21	55	M	3	—	GW	0	3	4	3.45	1.47
11	F	3	—	MIX	1	4	2	1.88	1.73	56	M	3	—	MIX	0	2	4	3.93	4.32
12	F	1	—	MIX	1	4	2	1.66	1.74	57	F	2	—	GW	1	6	2		
13	F	3	—	GW	2	1	3	2.78	3.06	58	F	1	—	GW	2	6	6		
14	F	1	—	GW	2	4	3	2.40	0.80	59	F	1	+	GW	1	4	6		
15	M	2	+	GW	0	3	4	3.69	1.42	60	F	1	+	GW	1	3	5		
16	M	3	—	GW	1	2	3	3.12	1.57	61	F	2	—	GW	0	7	5		
17	F	3	+	MIX	0	4	3	0.94	1.28	62	F	2	—	GW	1	1	3		
18	M	1	—	SW	2	4	3	4.39	2.72	63	M	2	—	GW	1	5	5		
19	M	3	+	SW	2	5	4	2.99	0.64	64	F	1	—	GW	1	5	4		
20	F	3	+	GW	2	2	5	2.76	0.97	65	F	1	+	GW	2	5	5		
21	M	2	+	SW	2	4	1	2.24	0.13	66	F	1	+	GW	1	4	5		
22	M	1	—	MIX	1	3	3	2.76	2.45	67	F	1	—	GW	1	3	4		
23	M	3	—	GW	1	3	2	2.77	2.08	68	F	3	—	MIX	2	4	3		
24	F	2	—	GW	0	6	3	2.55	1.45	69	F	2	—	MIX	2	6	5		
25	F	1	+	GW	2	4	3	3.21	0.89	70	M	3	—	GW	2	6	4		
26	F	1	+	MIX	2	4	2	3.53	2.66	71	F	1	+	GW	1	6	4		
27	M	1	+	GW	2	4	4	2.79	3.02	72	F	1	+	GW	0	6	5		
28	F	2	+	GW	1	4	3	2.98	1.68	73	M	1	+	GW	0	4	5		
29	F	3	—	GW	2	4	5	1.62	2.37	74	F	2	—	GW	2	6	5		
30	F	1	—	GW	1	2	5	3.51	1.74	75	M	1	—	GW	2	2	6		
31	F	3	—	GW	0	4	3	1.86	1.23	76	M	2	+	GW	2	4	5		
32	F	3	—	GW	0	1	1	2.92	2.65	77	M	1	+	GW	0	3	2		
33	M	2	+	GW	1	4	4	2.51	1.38	78	F	1	+	GW	2	4	4		
34	M	2	+	GW	0	5	4	2.26	1.26	79	F	1	+	GW	1	4	3		
35	M	1	—	SW	2	1	3	1.83	1.33	80	F	1	+	GW	0	4	4		
36	M	1	+	MIX	1	4	4	3.19	3.00	81	F	1	+	GW	2	4	5		
37	M	1	—	GW	1	4	5	2.79	3.02	82	M	1	—	GW	2	5	5		
38	F	3	—	GW	0	5	6	2.04	0.82	83	M	2	+	GW	2	2	6		
39	M	1	—	SW	2	6	5	1.51	3.06	84	M	1	—	GW	2	4	2		
40	F	2	—	SW	2	5	4	0.85	2.12	85	F	1	—	GW	2	5	6		
41	F	3	—	SW	2	5	3	2.88	1.26	86	M	1	—	GW	1	5	6		
42	M	2	+	GW	0	4	5	2.94	1.98	87	F	3	—	SW	0	7	5		
43	F	2	—	GW	1	4	3	3.27	1.65	88	F	3	—	SW	2	7	5		
44	M	3	—	GW	2	1	5	3.50	3.54	89	F	3	—	SW	2	6	5		
45	F	2	—	SW	2	5	4	1.71	1.35	90	M	3	+	SW	1	4	5		

person nr.	sex	age	smoking	source	community size	ISO-selection	DCB-selection	person nr.	sex	age	smoking	source	community	ISO-selection	DCB-selection
91	M	3	+	SW	1	7	7	120	F	3	—	SW	1	6	3
92	F	2	+	SW	2	6	1	121	F	1	+	GW	0	6	5
93	M	1	—	SW	1	7	1	122	F	3	—	SW	2	6	5
94	F	3	—	SW	1	5	4	123	F	1	—	SW	2	3	2
95	F	3	+	SW	0	4	4	124	F	3	—	MIX	1	7	6
96	M	3	+	SW	1	6	5	125	M	2	—	GW	2	6	5
97	F	2	+	SW	0	7	4	126	F	1	+	GW	0	6	5
98	M	2	—	SW	1	5	4	127	F	2	+	GW	0	5	6
99	M	3	—	SW	2	6	4	128	F	3	—	GW	2	6	4
100	M	3	+	SW	2	6	5	129	F	1	—	GW	2	3	4
101	F	1	+	SW	0	5	7	130	M	2	—	GW	2	7	6
102	M	3	—	GW	1	2	4	131	F	1	+	GW	2	3	2
103	M	1	+	SW	2	4	4	132	F	2	—	GW	2	7	4
104	F	2	+	SW	2	6	6	133	M	1	—	GW	2	3	5
105	F	1	+	SW	2	6	5	134	M	1	—	GW	2	3	5
106	M	1	+	SW	2	7	7	135	F	1	—	GW	1	3	4
107	M	1	+	SW	2	5	6	136	F	1	+	GW	1	6	6
108	F	1	—	SW	2	1	6	137	M	2	+	GW	1	6	5
109	F	1	—	SW	2	3	4	138	F	1	+	GW	0	4	4
110	M	1	+	SW	2	6	2	139	M	3	+	GW	0	7	5
111	F	3	—	SW	2	7	6	140	M	1	+	GW	1	5	5
112	M	3	—	SW	2	3	4	141	F	3	+	GW	2	3	5
113	F	1	+	SW	2	6	5	142	F	1	+	GW	1	6	2
114	M	1	—	SW	2	4	3	143	M	1	+	GW	1	4	
115	F	3	—	SW	2	5	5	144	F	2	+	GW	0	4	6
116	M	3	+	SW	2	7	4	145	F	2	—	GW	0	6	7
117	M	1	+	SW	2	3	6	146	M	2	+	GW	2	7	6
118	F	1	+	SW	2	4	6	147	F	3	—	GW	2	6	7
119	M	1	—	SW	2	6	5	148	F	2	+	GW	0	6	4

Table 3.A.1: Calculated t-values for the association between individuel characteristics and the odour sensitivity for aqueous solutions of isoborneol and o-dichlorobenzene.

Parameter	t-values*			
	Selection data		Panel data	
	ISO	DCB	ISO	DCB
Sex	1.81	−0.77	1.77	0.69
Age	1.61	0.73	1.21	0.87
Smoking	1.45	1.85	−0.81	1.23
Source	2.14	−1.15	2.66	1.65
Community size	−0.17	1.25	−0.31	−0.21

*$|t| > 1.96$ for $\alpha < 0.05$

Correlation between OTC for isoborneol and o-dichlorobenzene

The coefficient of correlation of the listed average OTC values for isoborneol and o-dichlorobenzene relating to the panel members amounts to 0.48 (t=3.97). This means that the odour sensitivity of panel members for one compound is significantly associated with the odour sensitivity for the other compound.

Appendix 3.2

Correction procedure for changes in the use of a taste rating scale by a panel

As stated in paragraph 3.5.2 a correction procedure had to be applied for a changed use by the panel members of the taste rating scale during the second and third session as compared to the first session. The correction method which was worked out in co-operation with Mr. G. de Graan is based on estimating the tendency of panel members during the first session to use the first items of the scale more frequently and the higher items less frequently. Furthermore this correction method had to meet with the constraint of the score-balance for the total number of taste scores for a certain type of water during each session.

If it is supposed that all panel members show the same changed attitude towards the taste scale for all types of drinking water, and that these changes concern the use of only the next higher category of the taste scale, it can be stated that n scores in the first category of the taste scale for a certain type of water are comparable to n_1 scores in the same scale category during the second and third session, or:

$$n_1' = \beta \cdot \alpha_1 n_1$$

in which:

n_1' = corrected number of scores for a certain type of water in the first class of the taste scale during the first session,

n_1 = original number of scores for a certain type of water in the first class of the taste scale during the first session,

α_1 = correction factor, with a value between 0-1, indicating the reduction in the use of the first class of the taste scale by the panel during the second and third session,

β = score-balance correction factor, necessary to produce comparable data for all sessions, amounting in this case to:

$$\frac{\text{sum of scores in 2nd and 3rd session}}{\text{sum of scores in 1st session}} = \frac{1019+520}{1074} = 1.43$$

In the same way it can be formulated for the other categories of the scale that:

$$n_2' = \beta[(1-\alpha_1)n_1 + \alpha_2 n_2]$$
$$n_3' = \beta[(1-\alpha_2)n_2 + \alpha_3 n_3]$$
$$n_4' = \beta[(1-\alpha_3)n_3 + \alpha_4 n_4]$$
$$n_5' = \beta[(1-\alpha_4)n_4 + n_5] \qquad (\alpha_5 = 1)$$

From these equations the values of α can be derived.

For the calculation of the average taste scale value (T) of a certain type of drinking water the following equation applies:

$$T = (f_1 \cdot n_1 + f_2 \cdot n_2 + f_3 \cdot n_3 + f_4 \cdot n_4 + f_5 \cdot n_5) / (n_1 + n_2 + n_3 + n_4 + n_5);$$

in which f is the scale value of the considered taste scale category, or:

$f_1 = 0$; $f_2 = 0.74$; $f_3 = 1.41$; $f_4 = 2.07$ and $f_5 = 2.87$

as described in section 3.3.

After application of the correction procedure for the data of the first session it can be calculated that:

$$T' = \frac{\sum_{i=1}^{5} f_i \cdot n_i'}{\sum_{i=1}^{5} n_i'} = \frac{\sum_{i=1}^{5} f_i \cdot n_i}{\sum_{i=1}^{5} n_i}$$

which implies that the correction for a changed use of the taste scale is mathematically equivalent to changed values for the items of the taste scale.

The latter equation implies that:

$f_1' = \alpha_1 \cdot f_1 (1 - \alpha_1) \cdot f_2$
$f_2' = \alpha_2 \cdot f_2 (1 - \alpha_2) \cdot f_3$
etc.
$f_5' = f_5$

Based on a changed use of the taste scale categories the corresponding values for α and f' have been calculated as presented in table 3.A.2.

Table 3.A.2: Deviation in the use of different taste scale categories of a drinking water derived from ground water during the first session compared to the second and third session and calculated values for α_i and f_i'

Taste scale category (i)	1	2	3	4	5
Number of scores in a category during 1st session (n_i)	111	567	278	90	28
Number of scores in a category during 2nd and 3rd session (n_i')	93	682	462	234	68
Values for α_i	0.58	0.76	0.67	0.78	1.0
Values for f_i	0	0.74	1.41	2.07	2.87
Values for f_i'	0.31	0.90	1.63	2.24	2.87

By using the f_i values for the scores in the five categories obtained during the second and third session and the f_i' values for the scores on the category scale for the data obtained during the first session, the mean taste scale values can be calculated and compared among each other for the total of the three sessions.

The values for α_1 and α_3 indicate that the panel members used the first and third item on the taste scale less frequently during the second and third session. After gaining experience the tendency to use the *"it tastes good"* item decreased in particular.

Appendix 5.1

List of organic compounds identified in 20 tapwaters in The Netherlands in 1976

Table 1: Hydrocarbons — Alkanes, Alkenes, Alkyl benzenes

Water type number	Cyclo-hexane C_6H_{12}	C_7 C_7H_{16}	C_8 C_8H_{18}	C_9 C_9H_{20}	C_{10} $C_{10}H_{22}$	C_{11} $C_{11}H_{24}$	C_{12} $C_{12}H_{26}$	C_{13} $C_{13}H_{28}$	C_{14} $C_{14}H_{30}$	C_8 C_8H_{16}	C_9 C_9H_{18}	C_{11} $C_{11}H_{22}$	Benzene C_6H_6	Toluene C_7H_8	o-Xylene C_8H_{10}	m/p-Xylene C_8H_{10}	Ethyl-benzene C_8H_{10}
1				0.005	0.03									0.03	0.005	0.01	0.01
2	0.1			0.01	0.01									0.01	0.005	0.01	0.03
3	0.03		0.005	0.005	0.005	0.005							0.03	0.03	0.005	0.03	0.006
4				0.01	0.01	0.01								0.01	0.005	0.01	0.005
5	0.1			0.01	0.03	0.03	0.01							0.1	0.01	0.01	0.03
6	0.03		0.01	0.03	0.03	0.03							0.03	0.03	0.005	0.01	0.005
7			0.03	0.03	0.3	0.03	0.01						0.03	0.03	0.01	0.03	0.03
8				0.01										0.01	0.01	0.01	
9				0.01	0.01	0.005								0.01		0.005	
10			0.005	0.01	0.1									0.005			
11					0.01									0.03	0.01	0.03	0.005
12	0.03			0.01	0.01	0.03								0.03	0.005	0.01	0.005
13			0.03	0.1	0.3	0.03	0.01	0.03	0.01					0.03	0.01	0.01	0.005
14				0.03	0.03	0.03							0.1	0.3	0.01	0.01	0.005
15			0.03	0.3	0.1								0.03	0.03	0.01	0.03	0.03
16				0.005	0.01									0.1	0.01	0.03	0.01
17			0.1	0.1	0.3	0.03	0.03							0.1	0.01	0.01	0.005
18					0.005	0.005	0.01							0.01	0.01	0.01	0.005
19		0.005	0.01	0.03	0.1	0.01				0.03	0.01	0.01	0.1	0.03	0.01	0.03	0.01
20	0.1				0.01	0.03	0.01							0.03	0.03	0.1	0.03
Detection limit and technique	0.005 GC/MS	0.005 GC/MS	0.005 GC/MS	0.005 GC/MS	0.005 GC/MS	0.005 GC/MS	0.005 GC/MS	0.005 GC/MS	0.005 GC/MS	0.005 GC/MS	0.005 GC/MS	0.005 GC/MS	0.01 GC/MS	0.005 GC/MS	0.005 GC/MS	0.005 GC/MS	0.005 GC/MS
Remarks			several isomers found	several isomers found	several isomers found	several isomers found	several isomers found						often solvent peak masked this compound				

Table 2: Hydrocarbons — Alkyl benzenes, Styrenes, Divinylbenzene, Methylindene, etc., Naphthalenes, Acenaphthene, Biphenyls

Water type number	C_3-benzenes C_9H_{12}	C_4-benzenes $C_{10}H_{14}$	Styrene C_8H_8	Methyl-styrenes C_9H_{10}	Ethyl-styrenes $C_{10}H_{12}$	Divinyl-benzene $C_{10}H_{10}$	Methyl-indene $C_{10}H_{10}$	4-Isopropenyl-1-methyl-cyclohexene $C_{10}H_{16}$	Dimethyl-4-vinyl-cyclohexene $C_{10}H_{16}$	Naphthalene $C_{10}H_8$	1-Methyl-naphthalene $C_{11}H_{10}$	2-Methyl-naphthalene $C_{11}H_{10}$	Dimethyl-naphthalene $C_{12}H_{12}$	Ethyl-naphthalene $C_{12}H_{12}$	Acenaphthene $C_{12}H_{10}$	Biphenyl $C_{12}H_{10}$	4-Methyl-biphenyl $C_{13}H_{12}$
1	0.03	0.005			0.03	0.01		0.005		0.03		0.005			0.005	0.03	0.03
2																	
3	0.03	0.005		0.005												0.03	
4	0.01	0.005				0.005		0.005		0.03	0.005	0.005			0.005	0.03	
5	0.01	0.005			0.005												
6	0.03	0.03		0.005	0.005	0.005		0.03		0.03	0.01	0.1			0.03		
7	0.01	0.005								0.005							
8	0.005																
9	0.005							0.005									
10																	
11	0.03			0.005	0.005			0.005		0.03	0.03		0.03	0.01	0.03	0.1	
12	0.01	0.005								0.03		0.005			0.01	0.01	
13	0.1	0.03			0.03	0.005				0.1		0.005				0.01	
14	0.03	0.005		0.01						0.1							
15	0.02									0.01							
16	0.02			0.005						0.1					0.01		
17	0.02									0.1					0.03	0.1	
18	0.02	0.005								0.1	0.01	0.03			0.01		
19	0.05		0.03		0.03		0.03	0.005		0.03		0.03				0.01	
20	1.0	0.1								0.1							
Detection limit and technique	0.005 GC/MS	0.005 GC/MS	0.005 GC/MS	0.005 GC/MS	0.005 GC/MS	0.005 GC/MS	0.005 GC/MS	0.005 GC/MS	0.005 GC/MS	0.005 GC/MS	0.005 GC/MS	0.005 GC/MS	0.005 GC/MS	0.005 GC/MS	0.005 GC/MS	0.005 GC/MS	0.005 GC/MS
Remarks	several isomers found	several isomers found	several isomers found	several isomers found	several isomers found	several isomers found											

Table 3: Hydrocarbons — Biphenyls, Fluorene, Anthracene, Phenanthrene, Pyrene, Dibenzo- etc.

Water type number	Biphenyls 2-Methyl-biphenyl $C_{13}H_{12}$	Diphenyl-methane $C_{13}H_{12}$	Fluorene $C_{13}H_9$	Anthracene $C_{14}H_{10}$	Phenanthrene $C_{14}H_{10}$	Methylene[def]phenanthrene $C_{14}H_{10}$	Methyl-phenanthrene $C_{15}H_{10}$	Pyrene $C_{16}H_{10}$	Dibenzo-hepta-fulvene $C_{16}H_{10}$	3-Phenyl-1,1,3-trimethyl-indane $C_{18}H_{20}$	3,4-Benzo-pyrene $C_{20}H_{12}$	3,4-Benzo-fluor-anthene $C_{20}H_{12}$	11,12-Benzo-fluor-anthene $C_{20}H_{12}$	2,3-Phenyl-ene-pyrene $C_{22}H_{12}$	1,12-Benzo-pery-lene $C_{22}H_{12}$
1	0.005			0.005	0.01	0.01	0.01		0.005						
2															
3															
4	0.005		0.03	0.05		0.1	0.03		0.3	0.01				0.005	0.005
5														0.005	
6				0.03		0.03					0.01			0.01	0.005
7				0.005							0.005				0.005
8				0.01											
9				0.02								0.01	0.01	0.01	0.01
10				0.005											
11	0.03	0.03	0.03	0.02		0.1			0.03						
12	0.005	0.005	0.03	0.01		0.01			0.01		0.01	0.01	0.01	0.005	
13				0.01		0.03			0.03					0.01	
14				0.005											
15				0.03					0.03						
16				0.02		0.03					0.005		0.01		0.005
17				0.02		0.01						0.01		0.005	
18	0.03	0.01	0.1	0.02	0.03	0.3	0.03	0.01	0.03		0.005	0.005			0.01
19				0.02		0.1		0.03	0.01						
20															
Detection limit and technique	0.005 GC/MS	0.005 GC/MS	0.005 GC/MS	0.005 GC/MS TLC	0.005 GC/MS	0.005 GC/MS	0.005 GC/MS	0.005 GC/MS	0.005 GC/MS	0.005 GC/MS	0.005 TLC	0.005 TLC	0.005 TLC	0.005 TLC	0.005 TLC
Remarks									most probable compound	most probable compound					

All concentrations in microgram/litre

COMPOUNDS CONTAINING OXYGEN

Table 1: Alcohols

Water type number	Alkanols C_5 $C_5H_{12}O$	Alkanols C_6 $C_6H_{14}O$	Alkanols C_7 $C_7H_{16}O$	Alkanols C_8 $C_8H_{18}O$	Alkanols C_9 $C_9H_{20}O$	Alkanols C_{11} $C_{11}H_{24}O$	trans-3-hexen-1-ol $C_6H_{12}O$	trans-2-hepten-1-ol $C_7H_{14}O$	trans-2-nonen-1-ol $C_9H_{18}O$	3,6-Dioxa-octanol-1 $C_6H_{14}O_3$	4-Propoxy-phenol $C_9H_{12}O_2$	1-Nonen-3-ol $C_9H_{18}O$	Di-butoxy-methanol $C_9H_{20}O_3$	3,7-Di-methyl-1,6-octa-dien-3-ol $C_{10}H_{18}O$	p-Menth-1-en-8-ol $C_{10}H_{18}O$	2-Methyl-isoborneol $C_{11}H_{20}O$	Geos-min $C_{12}H_{22}O$
1		0.1	0.03	0.01	0.03		0.01										
2					0.01												
3	0.03	0.03															
4					0.1												
5			0.03		0.03									0.005	0.03	0.005	
6			0.03										0.005				
7																	
8				0.01													
9				0.01													
10				0.005													
11		0.1	0.03	0.005	0.03												
12	0.1	0.1	0.01	0.01			0.1									0.03	
13	0.03		0.1	1.0					0.01								0.03
14					0.03						0.01		0.005				0.01
15										0.01							
16					0.3												
17																0.03	
18		0.01	0.03	3.0	0.005												0.03 0.01
19	0.3				0.1												
20																	
Detection limit and technique	0.005 GC/MS	0.005 GC/MS	0.005 GC/MS	0.005 GC/MS	0.005 GC/MS	0.005 GC/MS	0.005 GC/MS	0.005 GC/MS	0.005 GC/MS	0.005 GC/MS	0.005 GC/MS	0.005 GC/MS	0.005 GC/MS	0.005 GC/MS	0.005 GC/MS	0.005 GC/MS	0.005 GC/MS
Remarks		several isomers found		several isomers found	several isomers found		most probable compound			most probable compound	most probable compound	most probable compound				confirmed in odoro-gram	confirmed in odoro-gram

Table 2: Alcohols, Aldehydes, Ketones

Water type number	Alcohols Nonyl-Phenol $C_{15}H_{24}O$	Alkanals C_6 $C_6H_{12}O$	Alkanals C_7 $C_7H_{14}O$	Alkanals C_8 $C_8H_{18}O$	Alkanals C_9 $C_9H_{18}O$	Alkanals C_{10} $C_{10}H_{20}O$	Alkanals C_{11} $C_{11}H_{22}O$	Benz-aldehyde C_7H_8O	Dimethyl-benz-aldehyde $C_9H_{10}O$	Cinnam-aldehyde C_9H_8O	Ketones 2-Methyl-cyclobutan-one C_5H_8O	2-Cyclo-hexen-1-one C_6H_8O	4-Methyl-pentan-2-one $C_6H_{12}O$	Heptan-3-one $C_7H_{14}O$	4,4-Di-methyl-pentan-2-one $C_7H_{14}O$	6-Methyl-5-heptene-2-one $C_8H_{14}O$
1	0.03				0.005			0.1								
2																
3					0.005				0.03	0.1	0.005		1.0			0.005
4																
5					0.03	0.03			0.03							0.005
6					0.005											
7																
8																
9																
10			0.005		0.1			0.01	0.03	0.03			0.01			
11					0.005				0.03							
12					0.01			0.1	0.03	0.03				0.1		
13							0.03								0.1	
14							0.03									0.005
15			0.03	0.03	0.1	0.1		0.10	0.03			0.01				
16								0.1							0.01	0.01
17								0.01								
18		0.03	0.1		0.03											
19																
20																
Detection limit and technique	0.005 GC/MS	0.005 GC/MS	0.005 GC/MS	0.005 GC/MS	0.005 GC/MS	0.005 GC/MS	0.005 GC/MS	0.005 GC/MS	0.005 GC/MS	0.005 GC/MS	0.005 GC/MS	0.005 GC/MS	0.005 GC/MS	0.005 GC/MS	0.005 GC/MS	0.005 GC/MS
Remarks	most probable compound								most probable compound							

Table 3: Ketones

Water type number	Aceto-phenone C_8H_8O	2,4-Di-methyl-aceto-phenone $C_{10}H_{12}O$	3,4-Di-methoxy-aceto-phenone $C_{10}H_{12}O_3$	Bornan-2-one $C_{10}H_{16}O$	Dodec-2-ene-4-one $C_{12}H_{22}O$	Geranyl-acetone $C_{13}H_{22}O$	Di(tert)-butyl-1,4-benzo-quinone $C_{14}H_{10}O_2$	α,α,α,Triphenyl-aceto-phenone $C_{26}H_{20}O$
1	0.03							
2								
3								
4	0.1	0.005						
5							0.005	
6			0.005	0.01				
7								
8								
9								
10						0.005		
11								
12								
13	0.03	0.01					0.03	
14			0.01					
15				0.005			0.005	0.01
16			0.01					
17					0.03			
18							0.005	
19								
20								
Detection limit and technique	0.005 GC/MS	0.005 GC/MS	0.005 GC/MS	0.005 GC/MS	0.005 GC/MS	0.005 GC/MS	0.005 GC/MS	0.005 GC/MS
Remarks				most probable compound	most probable compound		most probable compound	

All concentrations in microgram/litre

COMPOUNDS CONTAINING OXYGEN — Ethers

Water type number	1,1-Di-methoxy-propane $C_5H_{12}O_2$	1,1-Di-methoxy-iso-butane $C_6H_{14}O_2$	1,3-Di-methoxy-iso-butane $C_6H_{14}O_3$	1,1-Di-ethoxy-ethane $C_6H_{14}O_2$	2,4,6-Tri-methyl-1,3,5-trioxane $C_6H_{12}O_3$	Bis-(2-methoxy-ethyl) ether $C_6H_{14}O_3$	2-Ethyl-hexyl-ether $C_8H_{14}O$	Di-butyl-ether $C_8H_{18}O$	Bis-(2-ethoxy-ethyl) ether $C_8H_{18}O_3$	Bis-(2-methoxy 2-ethoxy-ethyl) ether $C_{10}H_{22}O_5$	Di-cyclo-hexyl-ether $C_{12}H_{22}O$
1											
2		0.1									
3											
4		0.1						0.01			
5											
6		0.3						0.03			
7								0.03			
8		0.3	0.3					0.03			
9											
10	0.3	0.3									
11		0.1		0.005			0.01		0.03		
12		0.1									
13		0.1									
14	0.1	0.3									
15					0.03						
16		0.1									
17		0.1									0.005
18		0.1					0.005				
19						0.3				0.1	
20	0.1	0.3									
Detection limit and technique	0.01 GC/MS	0.01 GC/MS	0.01 GC/MS	0.005 GC/MS	0.005 GC/MS	0.005 GC/MS	0.005 GC/MS	0.005 GC/MS	0.005 GC/MS	0.005 GC/MS	0.005 GC/MS
Remarks										most probable compound	

COMPOUNDS CONTAINING OXYGEN — Esters — Acetates

Water type number	Ethyl-acetate $C_4H_8O_2$	Isopropyl-acetate $C_5H_{10}O_2$	2-Methoxy-ethyl-acetate $C_5H_{10}O_3$	Butyl-acetate $C_6H_{12}O_2$	2-Ethoxy-ethyl-acetate $C_6H_{12}O_3$	2,2-Dimethyl-propyl-acetate $C_7H_{14}O$	2-Ethyl-hexyl-acetate $C_{10}H_{20}O_2$	Linalyl-acetate $C_{12}H_{20}O_2$	(Cis-2-(cis-iso-propenyl)-3-methyl-cyclohexyl)-acetate $C_{12}H_{20}O_2$	Butyl-formiate $C_5H_{10}O_2$	Methyl-butyrate $C_5H_{10}O_2$	Methyl-iso-butyrate $C_5H_{10}O_2$	Methyl-pentanoate $C_6H_{12}O$	Methyl-2-methyl-butyrate $C_6H_{12}O_2$	Hexyl-butyrate $C_{10}H_{20}O_2$
1											0.1	0.1		0.01	
2		0.03								0.03	0.1	0.1		0.1	
3										0.1	0.1	0.1			
4												0.03			
5										0.1				0.3	
6								0.005			1.0			0.1	
7										0.1	0.3			0.1	
8															
9						0.03									
10											0.3				
11	0.03			0.005	0.005					0.1	0.1				
12				0.03											
13															
14					0.005						0.3	0.3			
15			0.1								0.3	0.3		0.1	
16								0.03							1.0
17											0.1	0.1			
18											0.1	0.1	0.03		
19								0.01	0.01			0.1			
20															
Detection limit and technique	0.005 GC/MS	0.005 GC/MS	0.005 GC/MS	0.005 GC/MS	0.005 GC/MS	0.005 GC/MS	0.005 GC/MS	0.005 GC/MS	0.005 GC/MS	0.005 GC/MS	0.005 GC/MS	0.005 GC/MS	0.005 GC/MS	0.005 GC/MS	0.005 GC/MS
Remarks														most probable compound	

COMPOUNDS CONTAINING OXYGEN — Esters — Phthalates

Water type number	Dimethyl-phthalate $C_{10}H_{10}O_4$	Diethyl-phthalate $C_{12}H_{14}O_4$	Dipropyl-phthalate $C_{14}H_{18}O_4$	Diisobutyl-phthalate $C_{16}H_{22}O_4$	Dibutyl-phthalate $C_{16}H_{22}O_4$
1	0.005	0.03		0.03	0.03
2					0.03
3					
4		0.005		0.01	0.1
5					
6		0.01		0.01	0.005
7		0.03			
8		0.005		0.005	0.005
9					0.005
10		0.01	0.005	0.01	0.03
11				0.01	0.01
12					0.03
13	0.03	0.1			0.1
14		0.01		0.01	0.01
15					0.3
16		0.005		0.005	0.005
17					0.03
18		0.01		0.03	0.1
19		0.03		0.01	0.03
20					0.03
Detection limit and technique	0.005 GC/MS	0.005 GC/MS	0.005 GC/MS	0.005 GC/MS	0.005 GC/MS
Remarks					

All concentrations in microgram/litre

COMPOUNDS CONTAINING HALOGEN

Haloforms

Water type number	Chloroform CHCl$_3$	Bromodichloromethane CHCl$_2$Br	Dichloroiodomethane CHCl$_2$I	Dibromochloromethane CHClBr$_2$	Bromoform CHBr$_3$	Bromochloroiodomethane CHClBrI	Tetrachloromethane CCl$_4$	1,2-Dichloroethane C$_2$H$_4$Cl$_2$	1,2-Dichloroethene C$_2$H$_2$Cl$_2$	Trichloroethene C$_2$HCl$_3$	Tetrachloroethene C$_2$Cl$_4$	Hexachloroethane C$_2$Cl$_6$	3-Bromopropyne C$_3$H$_3$Br	1,2-Dibromopropane C$_3$H$_6$Cl$_2$	Dibromopropadiene C$_3$H$_2$Br$_2$	Bromochlorobutene C$_4$H$_6$ClBr	Hexachlorobutadiene C$_4$Cl$_6$
1	0.5	0.01		0.01							0.03						
2																	
3	0.1						0.05				0.05						
4	0.8	0.1		0.01			0.01										
5							0.05										
6	2.0			0.01			0.1			9.0	0.1						
7	1.5	0.9		0.1			0.03				0.03						
8	0.1						0.05										
9										0.7							
10	0.1	0.1	0.03	0.3	0.3	0.01	0.03	0.03			0.1	0.03					0.01
11	1.5	1.0		0.8	1.0		0.1				0.1	0.06		0.1			0.03
12		0.01			0.01		0.05										0.1
13	10	10	0.3	5	1.0	0.03	0.03				3.0	0.1	0.03		0.03		
14	25	15	1.0	3	3.0	0.3					0.8	0.1	0.1	0.03	0.03		
15	0.3						0.05										
16	60	55		20	3.0		0.7					0.1					
17	0.2																
18	0.3	0.01		0.01			0.06				0.2	0.03	0.03		0.3		
19	40	35	0.1	10	10	0.1	0.3				0.2	0.1	0.03				
20	0.2						0.05			0.05	0.5						
Detection limit and technique	0.01 HS/ECD	0.01 HS/ECD	0.01 GC/MS	0.01 HS/ECD	0.01 GC/MS	0.01 HS/ECD	0.01 HS/ECD	0.01 HS/ECD	0.01 HS/ECD	0.01 HS/ECD	0.01 HS/ECD	0.005 GC/MS	0.01 GC/MS	0.01 GC/MS	0.01 GC/MS	0.01 GC/MS	0.01 GC/MS
Remarks													most probable compound		most probable compound	most probable compound	

COMPOUNDS CONTAINING HALOGEN

Halogenated hydrocarbons — Chlorobenzenes

Water type number	Chlorobenzene C$_6$H$_5$Cl	o-Dichlorobenzene C$_6$H$_4$Cl$_2$	m-Dichlorobenzene C$_6$H$_4$Cl$_2$	p-Dichlorobenzene C$_6$H$_4$Cl$_2$	1,2,4-Trichlorobenzene C$_6$H$_3$Cl$_3$	α-Hexachlorocyclohexane C$_6$H$_6$Cl$_6$	β-Hexachlorocyclohexane C$_6$H$_6$Cl$_6$	γ-Hexachlorocyclohexane C$_6$H$_6$Cl$_6$	Chloromethylbenzene C$_7$H$_7$Cl	p,p'-DDE C$_{14}$H$_8$Cl$_4$
1										
2										
3										
4		0.005		0.005						
5										
6										
7		0.005				0.01		0.01		
8										
9							0.01			0.01
10										
11	0.01	0.03	0.03	0.01	0.05	0.02		0.02		
12	0.005	0.03		0.03					0.005	
13	0.1	0.03						0.1		
14	0.03	0.01			0.005					
15		0.1								
16										
17	0.005	0.03	0.1	0.3	0.3					
18	0.005	0.03	0.01	0.1					0.005	
19	0.03	0.1	0.01	0.1	0.1	0.01		0.02		
20		0.01								
Detection limit and technique	0.005 GC/MS	0.005 GC/MS	0.005 GC/MS	0.005 GC/MS	0.005 GC/MS	0.01 EX/ECD	0.01 EX/ECD	0.01 EX/ECD	0.005 GC/MS	0.01 EX/ECD
Remarks										

COMPOUNDS CONTAINING HALOGEN

Chlorinated oxygen containing compounds

Water type number	1,1-Dichloroacetone C$_3$H$_4$Cl$_2$O	1,1,1-Trichloroacetone C$_3$H$_3$Cl$_3$O	Bis(2-chloroethyl) ether C$_4$H$_8$Cl$_2$O	Bis(2-chloroisopropyl) ether C$_6$H$_{12}$Cl$_2$O	Bis(dichloroisopropyl) ether C$_6$H$_{10}$Cl$_4$O	Bis(dichloro-n-propyl) ether C$_6$H$_{10}$Cl$_4$O	Toluoyl-chloride C$_7$H$_7$ClO	4-Chloro 2-ethyl phenol C$_8$H$_9$ClO
1								
2								
3								
4								
5								
6								
7								
8								
9								
10								
11				0.3			0.005	
12				0.1				
13								
14	1.0	0.01						
15	0.1			1.0				0.01
16				0.1	0.03	0.03		
17				1.0				
18			0.03	0.3		0.3		
19	0.1		0.03	0.3	0.3		0.03	
20			0.03	3.0				
Detection limit and technique	0.005 GC/MS	0.005 GC/MS	0.005 GC/MS	0.01 GC/MS	0.005 GC/MS	0.005 GC/MS	0.005 GC/MS	0.005 GC/MS
Remarks							most probable compound	most probable compound

All concentrations in microgram/litre

	COMPOUNDS CONTAINING NITROGEN											
	Nitrogen containing hydrocarbons						Nitrogen and oxygen containing compounds					
Water type number / Formula	2-Methyl-butane-nitrile	1-Hydro-benzthiazol	N-Ethyl-aniline	Methyl-quinoline	Nicotine	Diphenyl-amine	N,N'-Dime-thyl-N,N'-diphenyl-1,2-diamino-ethane	Nitro-benzene	Nitro-aniline	o-Nitro-toluene	o-Nitro-methoxy-benzene	2-Nitro-dimethyl-1,4-benzene-dicarboxyl-ate	N-Acetyl-N-Ethyl-aniline
	C_5H_9N	$C_6H_5N_3$	$C_8H_{11}N$	$C_{10}H_9N$	$C_{10}H_{14}N_2$	$C_{12}H_{11}N$	$C_{16}H_{20}N_2$	$C_6H_5NO_2$	$C_6H_6N_2O_2$	$C_7H_7NO_2$	$C_7H_7NO_3$	$C_{10}H_9NO_6$	$C_{10}H_{13}NO$
1													
2													
3			0.01										
4							0.005					0.03	
5													
6													
7													
8													
9													
10													
11								0.01	0.01	0.03			
12													
13			0.03		0.005								
14													
15													0.03
16	0.1											0.03	0.03
17													0.03
18					0.03							0.03	0.03
19	0.03	0.005			0.01			0.03		0.01	0.01	0.03	0.03
20							0.03						0.03
Detection limit and technique	0.005 GC/MS	0.005 GC/MS	0.005 GC/MS	0.005 GC/MS	0.005 GC/MS	0.005 GC/MS	0.005 GC/MS	0.005 GC/MS	0.005 GC/MS	0.005 GC/MS	0.005 GC/MS	0.005 GC/MS	0.005 GC/MS
Remarks	most probable compound					most probable compound		most probable compound					

	COMPOUNDS CONTAINING NITROGEN					
	Nitrogen and chlorine containing compounds			Nitrogen and sulphur containing compounds		
Water type number / Formula	2-Chloro-aniline	3,4-Dichlo-aniline	5-Chloro-o-toluidine	Ethyl iso-thiocyanate	Benzo-thiazole	2-(Methyl-thio)-benzo-thiazole
	C_6H_6NCl	$C_6H_5NCl_2$	C_7H_8NCl	C_3H_5NS	C_7H_5NS	$C_8H_7NS_2$
1						
2						
3						
4		0.03			0.005	
5						
6						
7						
8						
9						
10						
11					0.03	0.01
12					0.06	
13					0.01	
14						
15				0.01		0.01
16					0.01	
17	0.3	0.03	0.1			
18	0.01				0.01	
19						
20	0.3		0.3			
Detection limit and technique	0.005 GC/MS	0.005 GC/MS	0.005 GC/MS	0.005 GC/MS	0.005 GC/MS	0.005 GC/MS
Remarks	could also be other isomer					

	COMPOUNDS CONTAINING SULPHUR				MISCELLANEOUS						
Water type number / Formula	(Ethyl-thio-)-benzene	Dibenzo-thio-phene	2,5-Di-isobutyl-thio-phene	Di-phenyl-sul-fone	Tri-chloro-nitro-methane	Tri-ethyl-phos-phate	2-chloro-ethyl-nitro-phenyl-sulfone	Bis(diethyl-phosphoro-dithionate) methylene	Butyl-benzene-sulfon-amide	Tri-butyl-phos-phate	Cholin-esterase-inhibi-tors
	$C_8H_{10}S$	$C_{12}H_{10}S$	$C_{12}H_{20}S$	$C_{12}H_{10}SO_2$	CCl_3NO_2	$C_6H_{15}PO_4$	$C_8H_8ClNO_4S$	$C_9H_{22}P_2O_4S_4$	$C_{10}H_{15}NO_2S$	$C_{12}H_{27}PO_4$	
1											
2											
3											
4											
5											
6	0.01										
7											
8											
9											
10											
11				0.005		0.01		0.3	0.005		
12						0.01					
13			0.005		3.0	0.01					
14					3.0						
15			0.01		0.3	0.3					
16						0.1					
17						0.1		0.005			
18						0.03		0.005	0.01		
19		0.01		0.03		0.1			0.01	0.03	0.7
20						0.3					
Detection limit and technique	0.005 GC/MS	0.005 GC/MS	0.005 GC/MS	0.005 GC/MS	0.01 GC/MS	0.005 GC/MS	0.005 GC/MS	0.005 GC/MS	0.005 GC/MS	0.005 GC/MS	0.2
Remarks	most probable compound					most probable compound	most probable compound				as pera-oxon

All concentrations in microgram/litre

Appendix 6.1

OTC/LD$_{50}$ ratios for natural organic compounds in water

Number	Name	OTC (g/m³)	LD$_{50}$ (mg/kg) (oral)	Animal	Ratio OTC/LD$_{50}$
1	Acetaldehyde	0.015	1930	rat	$7.8 \cdot 10^{-6}$
2	Acetic acid	24	3310	rat	$7.2 \cdot 10^{-3}$
3	Benzaldehyde	0.035	1300	rat	$2.6 \cdot 10^{-5}$
4	Butanal	0.07	2490	rat	$2.8 \cdot 10^{-5}$
5	2,3-Butanedione	2.3	1580	rat	$1.5 \cdot 10^{-3}$
6	Butanoic acid	4.0	2940	rat	$1.4 \cdot 10^{-3}$
7	1-Butanol	2.8	2510	rat	$1.1 \cdot 10^{-3}$
8	2-Butanone	50	3100	rat	$1.6 \cdot 10^{-2}$
9	2-Butenal	0.53	300	rat	$1.8 \cdot 10^{-3}$
10	Coumarin	0.05	680	rat	$7.4 \cdot 10^{-6}$
11	Decanal	0.0001	3730	rat	$2.7 \cdot 10^{-8}$
12	Dimethylamine	23	698	rat	$3.3 \cdot 10^{-2}$
13	Dimethylsulfide	0.012	3300	rat	$3.6 \cdot 10^{-6}$
14	1.8-Epoxy-p-menthane	0.012	2480	rat	$4.8 \cdot 10^{-6}$
15	Ethanol	800	5560	guinea pig	$1.4 \cdot 10^{-1}$
16	2-Ethylbutanoic acid	10	2200	rat	$4.5 \cdot 10^{-3}$
17	2-Ethyl-1-Butanol	0.2	1850	rat	$1.1 \cdot 10^{-4}$
18	2-Ethyl-1-Hexanol	1.3	800	rat	$1.6 \cdot 10^{-3}$
19	2-(Ethylthio)ethanol	0.01	2320	rat	$4.3 \cdot 10^{-6}$
20	Formic acid	1500	1210	rat	$1.2 \cdot 10^{0}$
21	2-Furancarboxaldehyde	3.0	127	rat	$2.4 \cdot 10^{-2}$
22	Guaiacol	0.003	725	rat	$4.1 \cdot 10^{-6}$
23	Hexanal	0.03	4890	rat	$6.1 \cdot 10^{-6}$
24	Hexanoic acid	9	3000	rat	$3.0 \cdot 10^{-3}$
25	1-Hexanol	5.2	4590	rat	$1.1 \cdot 10^{-3}$
26	2-Hydroxyethylpropanoate	14	2580	mouse	$5.4 \cdot 10^{-3}$
27	Indole	0.3	1000	rat	$3.0 \cdot 10^{-4}$
28	β-Ionone	0.000007	4590	rat	$1.5 \cdot 10^{-9}$
29	Isopentylalcohol	7.0	3380	rat	$2.1 \cdot 10^{-3}$
30	Linalool	0.006	2790	rat	$2.2 \cdot 10^{-6}$
31	Methylsalicylate	0.1	887	rat	$1.1 \cdot 10^{-4}$
32	2-Mercaptoethanol	0.64	300	rat	$2.1 \cdot 10^{-3}$
33	Methanal	50	800	rat	$6.3 \cdot 10^{-2}$
34	3-Methylbutanol	0.25	3380	rat	$7.4 \cdot 10^{-5}$
35	2-Methylpropanoic acid	8.1	280	rat	$2.9 \cdot 10^{-2}$
36	Pentanal	0.012	4760	rat	$2.5 \cdot 10^{-6}$
37	Phenol	5.9	414	rat	$1.4 \cdot 10^{-2}$
38	Pivalic acid	50	500	rat	$1.0 \cdot 10^{-1}$
39	Propanoic acid	200	4290	rat	$4.7 \cdot 10^{-2}$
40	1-Propanol	9.0	1870	rat	$4.8 \cdot 10^{-3}$
41	Vanillin	4.0	1580	rat	$2.5 \cdot 10^{-3}$

Appendix 6.2

OTC/LD$_{50}$ ratios for industrial organic compounds in water

Number	Name	OTC (g/m³)	LD$_{50}$(mg/kg) (oral)	Animal	Ratio OTC/LD$_{50}$
1	Acrylonitrile	19	93	rat	$2.0 \cdot 10^{-1}$
2	Aldrin	0.017	39	rat	$3.9 \cdot 10^{-4}$
3	Aniline	70	440	rat	$1.6 \cdot 10^{-1}$
4	Benzene	10	3400	rat	$3.0 \cdot 10^{-3}$
5	Biphenyl	0.0005	2180	rat	$2.3 \cdot 10^{-7}$
6	Bis (2-chloroethyl)ether	0.36	75	rat	$4.8 \cdot 10^{-3}$
7	Bis (2-chloroisopropyl) ether	0.3	240	rat	$1.3 \cdot 10^{-3}$
8	Chlordane	0.0005	283	rat	$1.8 \cdot 10^{-6}$
9	Chlorobenzene	0.1	2910	rat	$3.4 \cdot 10^{-5}$
10	Chloroform	0.1	300	rat	$3.3 \cdot 10^{-4}$
11	2-Chlorophenol	0.0002	670	rat	$3.0 \cdot 10^{-7}$
12	2-Chlorotoluene	0.1	1231	rat	$8.1 \cdot 10^{-6}$
13	DDT	0.35	113	rat	$3.1 \cdot 10^{-3}$
14	1,4-Dichlorobenzene	0.03	500	rat	$6.0 \cdot 10^{-5}$
15	1,2-Dichloroethane	29	680	rat	$4.3 \cdot 10^{-2}$
16	2,4-Dichlorophenol	0.21	580	rat	$3.6 \cdot 10^{-4}$
17	2,6-Dichlorophenol	0.008	2940	rat	$2.7 \cdot 10^{-6}$
18	2,4-Dichlorophenoxyacetic acid	3.1	375	rat	$8.3 \cdot 10^{-3}$
19	Dichloropropane	0.0014	140	rat	$1.0 \cdot 10^{-5}$
20	Dieldrin	0.041	46	rat	$8.9 \cdot 10^{-4}$
21	Di-isobutylcarbinol	1.3	3560	rat	$3.7 \cdot 10^{-4}$
22	Dimethylparathion	0.01	4	rat	$2.5 \cdot 10^{-3}$
23	2,6-Dinitrotoluene	0.1	177	rat	$5.6 \cdot 10^{-4}$
24	Endrin	0.018	3	rat	$6.0 \cdot 10^{-3}$
25	Ethylbenzene	0.7	4930	rat	$1.4 \cdot 10^{-4}$
26	Ethylacrylate	0.007	830	rat	$8.4 \cdot 10^{-6}$
27	Ethylbenzene	0.2	3500	rat	$5.7 \cdot 10^{-5}$
28	Heptachlor	0.02	40	rat	$5.0 \cdot 10^{-4}$
29	α-HCH	0.09	500	rat	$1.8 \cdot 10^{-4}$
30	γ-HCH	12	76	rat	$1.6 \cdot 10^{-1}$
31	Isopropylester of 2,4-D	0.003	700	rat	$4.3 \cdot 10^{-6}$
32	Malathion	1.0	599	rat	$1.7 \cdot 10^{-3}$
33	2-Mercaptobenzothiazole	1.8	3000	rat	$6.0 \cdot 10^{-4}$
34	Methoxychlor	4.7	5000	rat	$9.4 \cdot 10^{-4}$
35	2-Methylnaphthalene	0.01	4360	rat	$2.3 \cdot 10^{-6}$
36	Napthalene	0.005	1780	rat	$2.8 \cdot 10^{-6}$
37	3-Nitrotoluene	0.13	1072	rat	$1.2 \cdot 10^{-3}$
38	4-Nitrotoluene	0.003	2144	rat	$1.4 \cdot 10^{-6}$
39	4-Nitrophenol	10	350	rat	$2.9 \cdot 10^{-2}$
40	Parathion	0.04	2	rat	$2.0 \cdot 10^{-2}$
41	Phenylphosphonothionic acid, 0-ethyl-0)(p-nitro-phenyl)ester (epn 300)	0.02	8	rat	$2.5 \cdot 10^{-3}$
42	Propenylguaethol	0.01	2400	rat	$4.2 \cdot 10^{-6}$

Number	Name	OTC(g/m³)	LD$_{50}$(mg/kg) (oral)	Animal	Ratio OTC/LD$_{50}$
43	Pyridine	0.82	891	rat	$9.2 \cdot 10^{-4}$
44	Rotenone	0.36	132	rat	$2.7 \cdot 10^{-3}$
45	1,2,4,5-Tetrachlorobenzene	0.13	1500	rat	$8.7 \cdot 10^{-4}$
46	Toluene	1.0	3000	rat	$3.3 \cdot 10^{-4}$
47	Toxaphene	0.14	60	rat	$2.3 \cdot 10^{-3}$
48	1,2,4-Trichlorobenzene	0.005	756	rat	$6.6 \cdot 10^{-6}$
49	Trichloroethene	0.5	4920	rat	$1.0 \cdot 10^{-4}$
50	2,4,5-Trichlorophenoxy-acetic acid	2.9	300	rat	$1.0 \cdot 10^{-2}$
51	2(2,4,5-Trichlorophenoxy) propionic acid	0.78	650	rat	$1.2 \cdot 10^{-3}$
52	Vinyl propionate	0.04	4760	rat	$8.5 \cdot 10^{-6}$